A Compendium of Alfalfa Diseases

Published by the American Phytopathological Society

Cooperating:

Science and Education Administration
U.S. Department of Agriculture

The College of Agriculture and the Agricultural
Experiment Station, Kansas State University

The College of Natural and Agricultural Sciences and
the Agricultural Experiment Station, University of California

The College of Agriculture and the Agricultural
Experiment Station, University of Minnesota

The American Phytopathological Society

Authors

Joseph H. Graham, Coordinator
Beltsville Agricultural Research Center
Agricultural Research
Science and Education Administration
U.S. Department of Agriculture, Beltsville, MD 20705
Present address: Waterman-Loomis Company
7625 Brown Bridge Road
Highland, MD 20777

Fred I. Frosheiser
Agricultural Research
Science and Education Administration
U.S. Department of Agriculture
Department of Plant Pathology
University of Minnesota, St. Paul, MN 55108

Donald L. Stuteville
Department of Plant Pathology
Kansas State University, Manhattan, KS 66506

Donald C. Erwin
Department of Plant Pathology
University of California, Riverside, CA 92521

Cover: J. H. Elgin, Jr., D. G. Gilchrist, Ernst Horber

Reprinted 1980

ISBN: 0-89054-026-8
Library of Congress Catalog Card Number: 79-88555

The American Phytopathological Society
3340 Pilot Knob Road, St. Paul, Minnesota 55121

Printed in the United States of America

Preface

The purpose of this Compendium is to aid plant pathologists and other agricultural workers in the diagnosis of alfalfa (lucerne) diseases. It is designed especially for extension or advisory plant pathologists who have access to a microscope and for pathologists who have limited experience with alfalfa diseases. The Compendium should be useful to plant disease consultants, advisers in state department of agriculture and federal regulatory agencies, agribusiness representatives, county agricultural agents and area specialists, plant pathology and vocational agriculture teachers, students, growers, and researchers in all countries where alfalfa is grown.

Diseases are arranged according to causal agents: abiotic agents and biotic agents—bacterial diseases, fungal diseases, viral and related diseases, nematodal diseases, parasitic flowering plants, and insect injury. In general, the fungal diseases are arranged according to the part of the plant affected, namely foliage, crown, and root.

An identification guide and an index are included to assist in the field or laboratory.

Suggested control measures are purposely general, stressing principles and cultural practices that are not likely to become obsolete and are applicable in various alfalfa-growing areas of the world. For specific information on resistant cultivars and other control measures, the reader should consult recent literature, extension or advisory plant pathologists, or informed seed suppliers.

Comments regarding the general usefulness or omissions in the Compendium are welcome. With your suggestions, future editions will increase in practicality and value for the diagnosis of alfalfa diseases.

The authors are indebted to M. C. Shurtleff for excellent mycological drawings and also for advice on various sections of the Compendium. Photographs and advice by T. O'Driscoll, USDA, are appreciated. We thank the Plant Physiology Institute Office, Beltsville Agricultural Research Center, for support and assistance during preparation of the manuscript.

Scientists who prepared specific sections are acknowledged at the end of the section. Specifically, we thank G. D. Griffin, E. W. Hanson, R. K. Howell, J. L. Kugler, K. T. Leath, George Semeniuk, R. E. Welty, and C. B. Willis.

The authors wish to thank the following individuals for reviewing the entire manuscript, making numerous valuable suggestions, and contributing illustrations:

R. B. Carroll
R. C. Close (New Zealand)
J. H. Elgin, Jr.
D. H. Hall
E. W. Hanson
I. C. Harvey (New Zealand)
E. J. Hawn (Canada)
R. B. Hine
K. T. Leath
F. L. Lukezic
Wm. D. McClellan
S. A. Ostazeski
R. E. Welty
R. D. Wilcoxson
C. B. Willis (Canada)

We also thank the following who cooperated by offering assistance and suggestions while reviewing specific parts of the Compendium and helped in other ways:

Bernard Alliot (France)
D. F. Beard
John Brown
Brian Hanger (Australia)
R. K. Howell
Ivor Isaac (Wales)
C. I. Kado
R. D. Lumsden
S. D. Lyda
D. C. MacLean
Deane Weber
W. G. Willis

Contents

Introduction

Alfalfa, *Medicago sativa,* also known as lucerne, is native to the Iranian region and was probably cultivated before recorded history. Throughout Asia and Europe and into Mexico, invading armies sowed alfalfa for their horses and cattle. George Washington was among the first to grow alfalfa in the eastern USA, but alfalfa was introduced even earlier in the western USA from Mexico.

Today alfalfa, the Queen of the Forages, is probably the world's most important forage crop. To retain and increase its value, plant pathologists and scientists in related disciplines must increase their efforts to identify, understand, and control the many diseases that affect this crop. Alfalfa is a highly heterozygous and autotetraploid perennial, and realizing its full production potential is indeed a difficult but worthy goal (Fig. 1).

Disease Development

Alfalfa frequently shows pathologic symptoms not associated with pathogenic or parasitic organisms. These abiotic or noninfectious diseases are caused by imbalances of soil nutrients or water, extreme soil acidity or alkalinity, extreme temperatures, air pollutants, pesticides, mechanical agents, and other factors. The severity and type of injury vary with the stage of plant maturity and plant parts involved. Symptoms caused by noninfectious agents can be confused with those caused by biotic agents. In addition, abiotic agents often influence the severity of diseases caused by biotic agents.

The extent and severity of biotic alfalfa diseases depend on the degree of compatibility between the host and pathogenic agent and on the influence of the air-soil

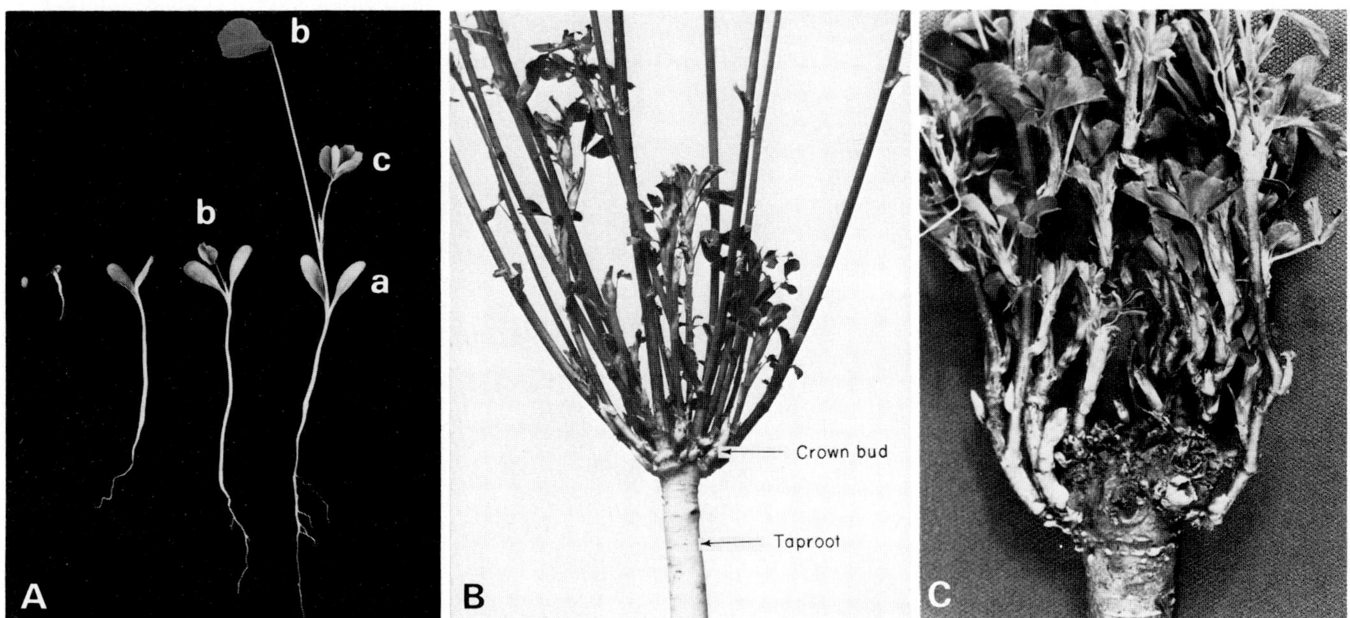

Figure 1. Growth stages of alfalfa. **A**, Development of a seedling: **a**, cotyledonary leaves; **b**, unifoliolate leaf; **c**, trifoliolate leaf. **B**, Crown of one-year-old plant uninjured by crown rotting pathogens. **C**, Crown of two-year-old plant. Note absence of crown buds in the center due to necrosis of tissue. This is the beginning of crown rot. (Courtesy D. C. Erwin)

environment on this association. Insects and other vectors are necessary to transmit some agents such as viruses (Fig. 2). By introducing toxic substances or by providing wounds for the pathogen to enter, nematode and insect feeding may predispose plants to infection by certain root-infecting or wilt-inducing agents.

Biotic Agents

Fungi

Fungi from at least 40 genera are by far the largest group of organisms attacking alfalfa. Fungi are simple, filamentous organisms without chlorophyll. Their spores germinate under favorable conditions to produce hyphae (filaments), and a mass of hyphae forms the mycelium or fungus body. Fungi produce spores by sexual or asexual means or both. Spores produced sexually tend to be thick-walled or contained in a protective structure and serve as a means of survival from one season to the next and as primary inoculum. Spores produced asexually, though more sensitive to unfavorable environmental conditions, are produced in large quantities and provide inoculum for secondary infections.

Plant pathogenic fungi penetrate the host plant either directly through the cuticle or indirectly through wounds or natural openings. They grow intercellularly or intracellularly or both in the plant. Some invade the vascular system and cause wilts. Nutritional requirements vary because pathogenic fungi range from those that can subsist indefinitely as saprophytes on plant refuse or in the soil, only occasionally invading living plants, to the obligate parasites that subsist only in living plants and cause rust, downy mildew, and powdery mildew. Some fungi can survive without a host for several years as resting bodies such as chlamydospores, sclerotia, or oospores. The occurrence and severity of a specific disease depend greatly on temperature and moisture conditions. Wind and water are the most important dispersing agents for fungus spores, but some spores may require carriers such as insects.

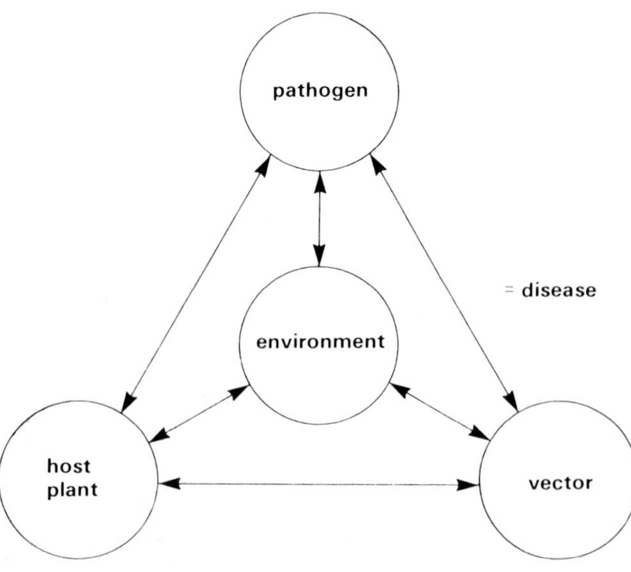

Figure 2. Disease triangle. Host, pathogen, and vector (transmission agent) interact, and a disease occurs on the host if the environment is favorable. (Reprinted, by permission, from A Compendium of Corn Diseases. ©1973, The American Phytopathological Society, Inc., St. Paul)

Bacteria

Bacteria are unicellular microorganisms that multiply by binary cell division. Most bacteria are saprophytes and play an important role in breaking down organic matter. Some are beneficial because they fix atmospheric nitrogen in the nodules on roots of legumes (Fig. 3).

The plant pathogenic bacteria are nonsporeforming rods. Most are motile and Gram-negative, and none is an obligate parasite. They enter plants only through wounds or natural openings such as stomates and hydathodes. A film of water on the host is required for penetration unless the bacteria are mechanically placed in the host by insects, nematodes, blowing soil particles, machinery, or other means. Some bacteria spread systemically through the vascular system; others cause local infections and spread by enzymatic action intercellularly or intracellularly or both. This enzymatic breakdown of cell structure in foliage causes a water-soaked symptom often present around the outer margin of active lesions.

Plant pathogenic bacteria survive from one season to another in plants, plant debris, seed, soil, or insects. They are spread by splashing and windblown rain, seed, hay, contaminated machinery, insects, or animals.

Viruses

These submicroscopic entities are strictly obligate parasites, replicating only in infected living cells. Morphologically, viruses may be long flexuous rods, polyhedral (spheres), or bacilliform in particle outline. Plant viruses consist of a nucleic acid core (generally RNA) protected by a protein coat. The manner of transmission varies among different viruses. They may be transmitted in nature through seed, by vectors such as insects, nematodes, fungi, or dodder, and by mechanical means. Viruses may persist in biennial or perennial plants, seeds, and vectors. They cause various symptoms including mosaic patterns, dwarfing of plants, and malformation of plant parts. Virus-infected

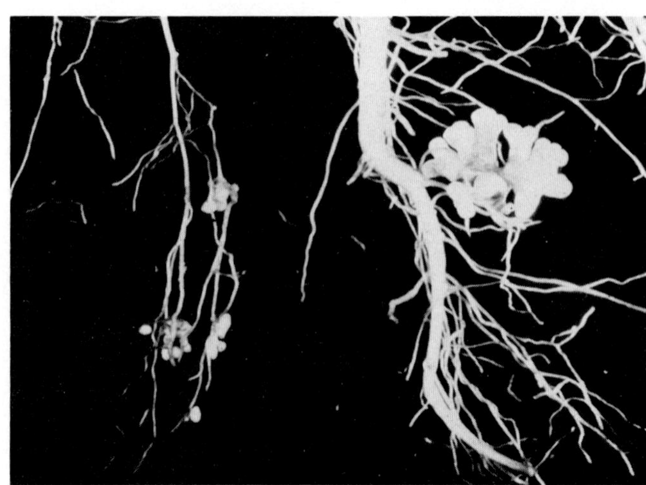

Figure 3. Ineffective (left) and effective (right) nodules on alfalfa roots. (Courtesy D. F. Weber, USDA)

alfalfa plants probably are prone to winter injury. Of numerous viruses isolated from alfalfa, alfalfa mosaic virus appears to be the most serious.

Until the mid 1960s, viruses were thought to cause many yellows-type diseases of plants, even though virus particles were never found. However, electron micrographs of thin sections of diseased phloem tissue and of some insect vectors reveal mycoplasmalike bodies not found in healthy tissue. These are submicroscopic, highly pleomorphic cells from 70 to 1,000 nm in diameter with unit membranes; they are inhibited by tetracycline antibiotics. They are transmitted by leafhoppers, dodder, and grafting but are not mechanically transmitted by sap. Some plant diseases once ascribed to viruses are caused by other agents including spiroplasmas, viroids, and bacteria.

Nematodes

Most nematodes (eelworms) are soil-inhabiting wormlike animals with elongated, unsegmented, semitransparent bodies that taper nearly to a point at both ends (Fig. 4). Plant parasitic nematodes reside in abundance in the root zone (upper 15 cm of soil) and are 15–35 μm wide and 0.3–4 mm long. All have a protrusile, hollow stylet (buccal spear) that is used to puncture plant cells (Fig. 5). They reproduce sexually, hermaphroditically, or parthenogenetically. Several species lack males. The life cycles include four larval stages, with the first molt usually occurring in the egg. A life cycle requires three to four weeks under optimum conditions.

Although some species attack the aerial and crown portions of plants, root damage is more common in alfalfa. Nematodes migrate a maximum of a few centimeters per year but are easily spread with soil particles moved by water, wind, or machinery. Long-distance dispersal occurs in nursery stock, seed, or debris in seed and hay and by other means.

Symptoms on roots may include knots or galls, excessive branching, injured (stubby) root tips, or a reduced root system. Wounds from nematode feeding also permit entry of bacteria and fungi that may eventually cause root rots. Root-knot galls are prone to infection by *Fusarium* and other root-rotting fungi. As a result of root injury, aboveground plant parts may show reduced growth, excessive wilting in hot dry weather, or nutrient deficiencies. Yield and quality of hay may be reduced. Some nematodes invade foliage and cause galls, abnormal development in foliar and floral parts, and necrotic lesions. Some nematodes also transmit plant viruses.

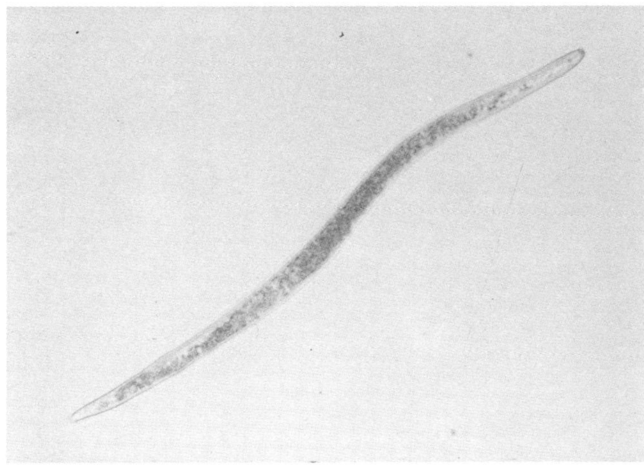

Figure 4. Lesion nematode (*Pratylenchus* sp.) (× 175). (Courtesy D. H. MacDonald)

Dodder

A parasitic flowering plant, dodder belongs to the *Cuscuta* genus in the Convolvulaceae family. Lacking chlorophyll, the plant is yellow instead of green and therefore cannot support itself except as a parasite.

Disease Control

The best way to reduce disease losses in alfalfa is to grow locally adapted disease-resistant cultivars. In northern areas, cultivars must be winter-hardy.

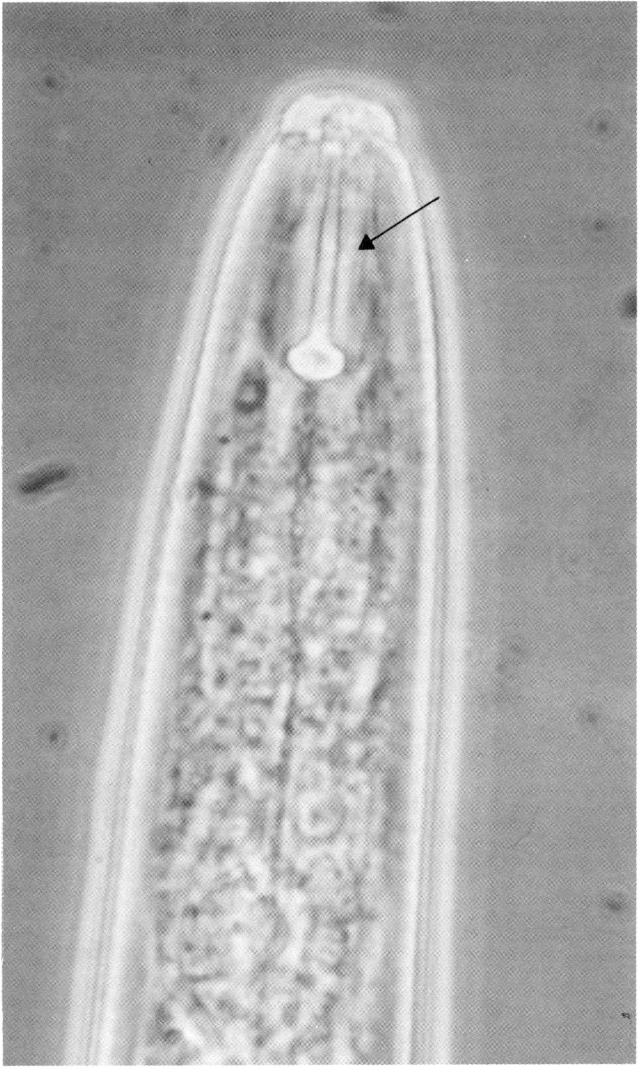

Figure 5. Anterior portion of a plant pathogenic nematode. Arrow points to retracted stylet (buccal spear), an identifying characteristic of plant pathogenic nematodes. (Courtesy D. H. MacDonald)

Adapted cultivars resistant to bacterial wilt are available. Cultivars resistant to one or more of the other diseases are available for specific areas. For example, anthracnose resistance has been bred into some of the mid-Atlantic cultivars and *Phytophthora* resistance has been incorporated in cultivars grown in irrigated and some nonirrigated areas. For information on new alfalfas with resistance to multiple diseases, consult local extension specialists and agricultural agents or advisers.

Good cultural and management practices help reduce disease losses. Shallow planting of high-quality seed in a firm, weed-free seedbed that is well drained, fertile, and at a suitable pH (6.5–8.0) is essential to obtain a thick and vigorous stand.

Harvest should be timed to maintain plant vigor, avoid leaf loss, and obtain high-quality forage. As a rule, first cutting in the spring should begin at the mid-to-late bud stage with later cuttings at first flower, which is usually at 30–40 day intervals. If foliar diseases begin to build up early, cutting earlier than normal can reduce leaf loss and reduce inoculum for the regrowth.

To minimize winter injury and maximize spring vigor in northern areas, alfalfa needs four to six weeks of favorable growing conditions in the fall to build up food reserves in the roots. Winter-injured plants are more vulnerable to crown and root diseases.

General References

BARNES, D. K., E. T. BINGHAM, R. P. MURPHY, O. J. HUNT, D. F. BEARD, W. H. SKRDLA, and L. R. TEUBER. 1977. Alfalfa germ plasm in the United States. Genetic vulnerability, use, improvement, and maintenance. U.S. Dep. Agric. ARS Tech. Bull. 1571. 21 p.

BARNES, D. K., J. H. ELGIN, Jr., F. I. FROSHEISER, E. L. SORENSEN, M. W. NEILSON, W. F. LEHMAN, K. T. LEATH, R. H. RATCLIFFE, and R. J. BUKER. 1974. Standard tests to characterize pest resistance in alfalfa varieties. U.S. Dep. Agric. ARS-NC-19. 23 p.

BOLTON, J. L. 1962. Alfalfa: Botany, cultivation, and utilization. Interscience Publishers, Inc., New York. 473 p.

BUCHANAN, R. E., and N. E. GIBBONS (eds.). 1974. Bergey's manual of determinative bacteriology, 8th ed. Williams & Wilkins Co., Baltimore, MD. 1,268 p.

DICKSON, J. G. 1956. Diseases of field crops. McGraw-Hill Book Co., Inc., New York. 517 p.

HANSON, C. H. (ed.). 1972. Alfalfa science and technology. Am. Soc. Agron., Inc., Madison, WI. 812 p.

O'ROURKE, C. D. 1976. Diseases of grasses and forage legumes in Ireland. An Foras Talũntais (The Agricultural Institute), Oak Park Research Centre, Carlow, Ireland. 115 p.

SAMPSON, K., and J. H. WESTON. 1954. Diseases of British grasses and herbage legumes. University Press, Cambridge. 118 p.

THORNE, G. 1961. Principles of nematology. McGraw-Hill Book Co., Inc., New York. 553 p.

Part I. Diseases Caused by Abiotic Agents

Nutrient Deficiencies and Toxicities

A lack or excess of essential plant nutrients leads to abnormal or reduced growth or both. When the mineral imbalance reaches a certain level, specific symptoms develop that may indicate the cause of the disease or disorder.

Potassium (K)

Potassium deficiency is perhaps the most common and the most easily recognized nutrient imbalance. Small white spots appear first at the margins of leaflets (Fig. 6). The areas between the spots turn yellow and die. In some instances the oldest leaves turn slightly to markedly yellow before a brown necrosis develops along the margins.

A second type of lesion consists of poorly defined gray spots on the upper surface at the distal margins on either side of the midrib of older leaves. These areas soon become a pinkish cinnamon color, and the resulting necrosis starts at the distal margin and gradually spreads across the entire leaf. Meanwhile, secondary shoots with small leaves may continue to be produced from the base of the affected plant.

Nitrogen (N)

Nitrogen deficiency, which develops in poorly nodulated plants, causes dwarfed, spindly growth with general yellowing (Plate 1). A pink color starts in the petioles of the oldest leaves and continues up the midrib on the underside of the leaflets. These leaves soon yellow and die, and the symptoms progress upward to successively younger leaves. Moderate deficiency slows growth and decreases branching, but it may not be recognized. Under favorable conditions *Rhizobium* nodule bacteria supply sufficient nitrogen for the plant.

Phosphorus (P)

Phosphorus deficiency symptoms often are not well defined. Roots are light brown and restricted in growth. Top growth is stunted, stiff, and erect. Leaves are small and abnormally dark green. The leaflets tend to fold together, and the undersides are purplish. Finally, the oldest leaves turn yellow and die.

Calcium (Ca)

Calcium deficiency symptoms are seldom seen in the field. However, in moist humid regions, high rates of lime are required for root development and nodulation and to supply nutrients essential for growth.

In young plants, the first symptom is sudden collapse of the petioles of the youngest fully developed leaves. The leaflets of the collapsed petioles remain turgid for two or more weeks, but the undersides of the leaflets become reddish purple. Meanwhile, marginal necrosis develops on young immature leaves, sometimes preceded by yellowing of the tissue. This necrosis commences at the leaflet tips and is initially blue-green

Figure 6. Potassium deficiency. (Courtesy Department of Agriculture, Victoria, Australia)

and later gray white. The affected margins curl up and in to form a characteristic funnel of the top of each leaflet (Plate 2).

In older plants that are growing rapidly and have several centimeters of woody stem, approximately the upper 2 cm of the stem may collapse suddenly (Fig. 7).

Magnesium (Mg)

Magnesium deficiency is more prevalent on acid, sandy soils and in wet seasons because magnesium is readily leached. Leaflets of deficient plants show interveinal chlorosis (Plate 3). The margins initially remain green but later become chlorotic, and the leaves die. These symptoms progress from older to younger leaves. In some plants the margins of the older leaves redden pronouncedly. The red area extends into the center of the lamina until only a small inverted V-shaped area of green tissue remains around the base of the midrib. Affected leaves finally die. The deficiency can be corrected by adding magnesium sulfate to the soil or by liming it with dolomitic limestone.

Iron (Fe)

Iron-deficiency symptoms rarely occur in the field. In some instances, however, iron may not be available in alkaline or calcareous soils. Excess applications of lime to acid soils, a high phosphorus content in neutral or alkaline soil, or excesses of zinc, copper, manganese, and other microelements may induce iron deficiency. In iron-deficient plants, chlorosis appears first between the veins of the youngest leaves, and the leaves turn progressively from yellowish green to a bleached yellow (Plate 4). Plants are stunted and occasionally the youngest leaves die.

Figure 7. Calcium deficiency causing sudden collapse of top of stem. (Courtesy Department of Agriculture, Victoria, Australia)

Boron (B)

Boron deficiency is common in many countries. The deficiency symptoms are most prominent during dry periods and, in fact, may be confused with drought symptoms or leafhopper yellowing. Plant tops become yellow and plants are dwarfed. Yellowing is followed by development of a purplish to rose-pink color on leaflet margins and on the undersides of the youngest fully developed leaves. Later the chlorosis extends in the interveinal tissues, and the upper leaf surface yellows and then turns red. The lower leaves remain a healthy green. Severe symptoms include shortening of the stem internodes, rosetting of the terminal leaves, and sometimes death of the growing point. Severely affected plants fail to flower; those less severely affected will flower but do not set seed (Plate 5).

Manganese (Mn)

Manganese deficiency is most evident in alkaline soils, especially those high in phosphorus. An excess of manganese may induce iron deficiency; conversely, an excess of iron may rarely induce manganese-deficiency symptoms. High levels of manganese also interfere with the uptake and utilization of molybdenum. Initial deficiency symptoms are a reduction in growth associated with a chlorosis that is most pronounced in the youngest leaves. The chlorosis occurs first in the interveinal tissues, while the veins remain green. Later, small light-brown necrotic areas, which are initially visible on the upper surface, develop in the chlorotic leaves (Plate 6).

Manganese toxicity can occur in acid soils. Older leaflets become chlorotic around the margins, and small necrotic spots may develop. In some soils (e.g., Richland soil at pH 5.2), adding lime to correct manganese toxicity can result in boron deficiency. In certain areas of Australia, manganese toxicity develops after heavy rains on acid soils (around pH 5) and also during extended hot, dry periods (Fig. 8).

Zinc (Zn)

Though reported in most alfalfa-growing areas of the world, zinc deficiency is rare. Addition of lime and phosphorus may reduce availability of zinc and cause

Figure 8. Manganese toxicity (left) and healthy plant (right). (Courtesy J. D. Brown, USDA)

deficiency symptoms. The first symptom is reduction in size and upward curling of the youngest leaves. Next, brown to bronze spots appear on the upper surface around the margins of the top (but not the youngest) leaves of the plant. When these bronzed areas later become white and necrotic, they also occur on the leaf undersurfaces. Some leaves have bronze spots distributed randomly over the upper surface of leaflets, rather than concentrated near edges.

The growth rate is soon retarded, tillering is reduced, and the entire plant becomes slightly chlorotic. New leaves, particularly those on tillers at the base of the plant, are smaller and stiffer than normal, and their margins tend to roll inward. The shape of the "little" leaves is also abnormal, and the apex of the leaflet may have a deep notch in the margin (Fig. 9). Finally a gray necrosis starts on the middle leaves of the plant at the tip of each leaflet and extends down the lamina until the entire leaf and top of the petiole are involved. These leaves tend to fall from the plant.

Copper (Cu)

The first symptom of copper deficiency is wilting of the younger leaves. A striking curvature or epinasty of the petioles of the younger leaves develops in some affected plants, causing the leaflets to bend back against the lower part of the petiole. Necrotic pale gray to white spots soon appear on the wilted leaves, either on the margins or more characteristically toward the base of each leaflet. Soon after the growing point dies, the leaflets and petioles die also (Fig. 10).

Figure 9. Zinc deficiency with characteristic "little leaf" symptom at base of plant. (Courtesy Department of Agriculture, Victoria, Australia)

Molybdenum (Mo)

Molybdenum deficiency occurs in soils rich in iron and in acid soils because molybdenum is less available at low pH. The element is essential for nodulation and assimilation of nitrogen in the plant; therefore, deficiency symptoms are similar to those of nitrogen deficiency. Also, if the metabolic demands of the plant are not met, leaflets may show interveinal white notching, followed by complete whitening of the tips and centers. Flowering usually is restricted.

Aluminum (Al)

Aluminum toxicity symptoms may develop on acid soils (below pH 5.5). Maximum aluminum concentration in the soil restricts the transport of phosphorus at pH 6 or lower. Affected plants are stunted and dark green. In high aluminum soils, roots are restricted and stubby. Chlorotic spots may develop on the lower leaves. Liming the soil eliminates the toxic effect of aluminum and increases the availability of phosphorus. When some high aluminum soils such as Bladen are limed, copper-deficiency symptoms may develop in the alfalfa plants.

Other Elements

Cobalt is thought to be essential for nodule-forming bacteria but has not proved essential for alfalfa.

Sulfur deficiency is becoming more prevalent, especially in light soils. The element is one of the organic compounds in plants, and deficiency symptoms are similar to those of nitrogen deficiency. Growth is stunted and leaves turn pale green or yellowish. One reason sulfur deficiency has increased is the switch from superphosphate to concentrated phosphate fertilizers that do not contain sulfur.

Chlorine is reportedly essential for growth, but deficiency symptoms have not been identified in the field.

Figure 10. Copper deficiency showing bending of petioles. (Courtesy Department of Agriculture, Victoria, Australia)

Correcting Nutrient Imbalances

By the time symptoms of a nutrient deficiency or toxicity are clearly recognizable, usually the crop is already damaged considerably. Adding the lacking major or trace element usually corrects the deficiency. Boron is formulated in fertilizers in some areas, and lime corrects toxicities such as aluminum and manganese. However, application of a nutrient, such as lime below the plow layer, is not always practical. Research shows that genotypes of crops can be selected for more efficient utilization of certain nutrients or for more tolerance to other nutrients.

Selected References

BAUR, K., G. A. HUBER, and L. C. WHEETING. 1941. Boron deficiency of alfalfa in Western Washington. Wash. Agric. Exp. Stn. Bull. 396. 16 p.

BERGER, K. C. 1958. Boron for alfalfa. Univ. Wis. Ext. Circ. 555. 6 p.

BOLTON, J. L. 1962. Plant nutrients, chemical composition, and nitrogen fixation. Ch. 9, pages 294-320 *in* Alfalfa: Botany, cultivation, and utilization. Interstate Publishers, Inc., New York. 473 p.

COLWELL, W. E., and C. LINCOLN. 1942. A comparison of boron deficiency symptoms and potato leafhopper injury on alfalfa. J. Am. Soc. Agron. 34:495-498.

MILLIKAN, C. R. 1958. Nutrient deficiency symptoms in lucerne and some pasture legumes. J. Dep. Agric. Victoria, Aust. 56:511-531.

NELSON, W. L., and S. A. BARBER. 1964. Nutrient deficiencies in legumes for grain and forage. Pages 143–180 *in* H. B. Sprague (ed.). Hunger signs in crops. David McKay Co., Inc., New York. 461 p.

WALKER, J. C. 1969. Plant Pathology. McGraw-Hill Book Co., Inc., New York. 819 p.

(Prepared in most part from material by C. R. Millikan)

Environmental Factors

Winter Injury

Winter injury to alfalfa may result from low temperatures and plant desiccation, particularly during periods without snow cover; prolonged coverage of the soil surface by ice sheets, which prevents normal exchange of gases; and alternate freezing and thawing, which mechanically injure roots and cause heaving.

Cold and desiccation may kill crown buds or injure them, resulting in weak shoots. New buds may form on the lower portions of the crown, and as the season progresses, recovery may be nearly normal. Thus, stands that appear unproductive early in the spring may produce normal yields later in the season. Phloem tissue in the roots, crowns, and crown buds is particularly susceptible to winter injury.

Plants weakened by a disease such as bacterial wilt are more susceptible to winterkill than are healthy plants. Conversely, winter injuries provide avenues of entrance for pathogens, making the plant more prone to disease. Death from effects of low-temperature pathogens early in the spring may be confused with winterkill. Allowing sufficient time for fall regrowth will allow replenishment of carbohydrate root reserves and will "harden off" the plants, thereby increasing their winterhardiness.

Figure 11. Heaving of alfalfa plants. (Courtesy J. H. Elgin, Jr., USDA)

Alfalfa cultivars differ greatly in winterhardiness and are generally classified as winter-hardy, moderately winter-hardy, and nonwinter-hardy. Generally, the higher the latitude, the hardier the cultivar must be. In some southerly areas, however, cultivars that do not break dormancy easily are required because several weeks of growing temperatures may occur during the winter, followed by temperatures much below freezing.

When ice sheets completely cover the plants, gas exchange is limited; the increased CO_2 concentration and depletion of O_2 lead to tissue decline and death. Less damage usually occurs when stubble protrudes through the ice, permitting gas exchange.

Heaving usually occurs on heavy, wet soils. Freezing that alternates with periods of thawing exerts an upward pressure on the crown. When the pressure is great enough, the crown and root are lifted. The taproot may break if the lower portion is held by soil that remains frozen solid (Fig. 11). Heaving is most severe on stands during the first winter following late summer or fall plantings because the root systems are short and poorly anchored. If many lateral roots break off, mechanical injury alone may kill the plant. Also, root injuries of partially heaved plants provide points of entry for the bacterial wilt and root-rotting pathogens. There is evidence that roots damaged by the clover root curculio *Sitona hispidula* (Fab.) and Phytophthora root rot are prone to heaving.

Selected References

FREYMAN, S., and V. C. BINK. 1967. Nature of ice-sheet injury to alfalfa. Agron. J. 59:557-560.

HEINRICHS, D. H. 1973. Winterhardiness of alfalfa cultivars in southern Saskatchewan. Can. J. Plant Sci. 53:773-777.

JONES, F. R. 1928. Winter injury of alfalfa. J. Agric. Res. 37:189-211.
MARK, J. J. 1936. The relation of reserves to cold resistance in alfalfa. Iowa Agric. Exp. Stn. Res. Bull. 208. 32 p.
WEIMER, J. L. 1930. Alfalfa root injuries resulting from freezing. J. Agric. Res. 40:121-143.

Frost and Late Spring Freeze Damage

After a period of mild weather in the spring, flush growth is tender and subject to freeze damage. In mild cases the top leaves appear to be water-soaked and a few die (Plate 7). A diagnostic symptom of frost injury is separation of the epidermis from the underlying tissue. This occurs in what appears to be uninjured tissue and becomes visible when the epidermis is teased away from the leaf or stem. However, entire shoots may be frozen and the first harvest delayed and reduced. Frost damage increases the plant's proneness to bacterial stem blight. The foliage of alfalfa cultivars differs in frost tolerance, but frost tolerance is not always associated with winterhardiness.

Scald or High-Temperature Flooding Injury

This problem is most serious in the hot desert valleys of the southwestern USA but also occurs to a limited extent in temperate regions where soil is saturated for long periods after irrigation or rainfall (Fig. 12). Water flowing over the hot soil in the desert valleys of California was thought to increase in temperature to a level capable of causing rapid death of the plants. Because the plants often die within three or four days after irrigation, growers called the phenomenon "scald." However, measurements indicated that water temperatures, even in the lower end of irrigation runs, were actually lower than ambient air temperatures. The disease is related to the combination of high soil and water temperatures and saturation of the soil with water. The critical factor is the length of time the soil is saturated at the high soil temperature.

Figure 12. Field in the Imperial Valley of California. The stand in the irrigation bed (foreground) was lost to excessive irrigation at high temperatures (scald). (Courtesy D. C. Erwin)

Symptoms

When air temperature exceeds 40–45°C, alfalfa becomes extremely susceptible to flooding injury. Foliage of all affected plants in the flooded area fades to light green and wilts even though the soil is wet. Roots of some plants completely rot and, when removed from the soil, have a putrid odor. The xylem elements of affected roots that are not rotted become brown and necrotic (Fig. 13). Recently mowed plants or those run over by a truck or tractor are much more susceptible than noninjured plants to flooding injury (Fig. 14).

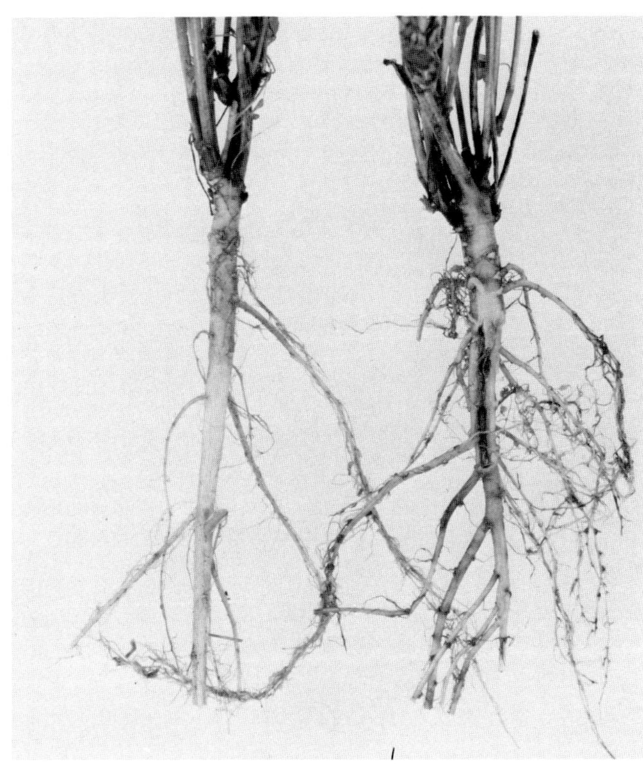

Figure 13. Plant at right is from saturated soil in a pot maintained at 39°C for 24 hr. Plant at left is from nonsaturated soil maintained at 39°C. Note vascular discoloration due to flooding at high temperature (scald). (Courtesy D. C. Erwin)

Figure 14. Effect of driving a tractor over an experimental plot of alfalfa before irrigating the plot heavily for 24 hr at high temperature (39°C). All plants in the wheel tracks were killed by scald. (Courtesy D. C. Erwin)

Cause

Experimental studies under controlled conditions show that high temperatures alone do not adversely affect alfalfa roots in unsaturated soil, but when the soil is saturated with water at high temperature, typical root and foliage symptoms occur. Because forcing air through water-saturated soil in pots prevents injury, lack of sufficient soil aeration probably is the main cause of vascular discoloration and root rot.

Control

1) Proper water management is the best control. Flooding for short periods (about 4 hr) when temperatures are high reduces the likelihood of scald; however, some soils remain saturated long after irrigation because of soil type, the slope of the land, and the length of the irrigated run.

2) Irrigation should be avoided when temperatures are excessively high (43–49°C).

3) Newly mowed plants should not be irrigated until enough regrowth occurs to prevent submersion of entire plants.

Selected References

ERWIN, D. C., B. W. KENNEDY, and W. F. LEHMAN. 1959. Xylem necrosis and root rot of alfalfa associated with excessive irrigation and high temperatures. Phytopathology 49:572-578.

ERWIN, D. C., and W. F. LEHMAN. 1974. Summer stand depletion of alfalfa in the low desert valleys of southern California. Pages 58-64. Fourth Calif. Alfalfa Symp.

LEHMAN, W. F., S. J. RICHARDS, D. C. ERWIN, and A. W. MARSH. 1968. Effect of irrigation treatments on alfalfa (Medicago sativa L.) production, persistence, and soil salinity in southern California. Hilgardia 39:277-295.

Air Pollution

Air pollutants are ubiquitous in nature. Sulfur dioxide (SO_2), hydrogen fluoride (HF), peroxyacetyl nitrate (PAN), and ozone are the major air pollutants known to injure alfalfa. They are taken up through stomates into leaf tissue. Only SO_2 and HF accumulate in tissue. HF and, to some extent, SO_2 can be diagnosed by analysis of leaf tissue. Ozone and PAN, considered to be the most widely dispersed photochemical pollutants, cause plant injury but break down completely in plant tissue.

The deposition and concentration of SO_2 and HF depend on the quantities of ore processed or fossil fuels burned (e.g., in steel mills), the concentrations of impurities in the ore or coal, and on meteorologic conditions. Generally, these pollutants follow the direction of prevailing winds, and the sphere of dispersion depends on the height of the stack. Photochemical air pollutants such as PAN and ozone result from the reaction of sunlight on oxides of nitrogen and hydrocarbons. High-temperature combustion generates oxides of nitrogen. Although both oxides of nitrogen and hydrocarbons may form at burning sources, a great proportion of these pollutants comes from automobile exhaust; thus, the greatest amount of pollution occurs near the most dense centers of population.

In areas of dense population and heavy inversion layers of air such as in the Los Angeles basin of California, air pollution drastically limits production, but in areas where cyclonic storms are frequent, damage from photochemical pollution is almost nonexistent.

Foliar symptoms on pinto bean, petunia plants, and some alfalfa cultivars can be used to differentiate ozone and PAN injury. SO_2 damage is specific enough to be diagnostic. Foliar injury is not always an adequate indicator of yield losses of alfalfa. One study showed that high dosages of HF, which induced foliar symptoms on alfalfa, did not reduce yields of the crop harvested immediately after treatment or in subsequent harvests.

Ozone

Typically, ozone causes some chlorosis, but the main symptom is appearance of small isolated necrotic areas to intense brown lesions distributed between the veins of injured leaves (Plate 8). If injury is substantial, leaves senesce and fall. Newly emerging leaves do not have symptoms. Ozone is considered to be the most widely distributed air pollutant that damages plants. Ozone damage to alfalfa in the field is sometimes confused with symptoms caused by nutritional disorders, pesticides, insects, diseases, and physiologic disorders. High concentrations of ozone generally are associated with low wind velocities, bright sunlight, and lack of cyclonic storms during the growing season.

Peroxyacetyl Nitrate (PAN)

Symptoms on leaves resemble those described for both ozone and SO_2 injuries. Symptoms may vary from small necrotic lesions to large necrotic leaf areas. In addition, a silver or copper sheen is frequently apparent on affected leaves and is attributed to the presence of PAN.

Sulfur Dioxide (SO_2)

Symptoms include chlorosis and narrow to broad, cream to light brown bands between veins and frequently at the leaf margins (Fig. 15). Leaves with 20–30% injury fall prematurely. Symptoms may resemble injury from certain herbicides.

Hydrogen Fluoride (HF)

Symptoms are similar to those induced by SO_2. For confirmation, check for nearby sources of HF and analyze leaf tissue for fluorides, since fluoride accumulates in plant tissue (Fig. 16).

Figure 15. Sulfur dioxide injury. (Courtesy R. K. Howell, USDA)

Figure 16. Hydrogen fluoride injury. Control (left); plant exposed to hydrogen fluoride fumigation (right). (Courtesy A. E. Hitchcock)

Control

1) Selection for resistance to the photochemical air pollutants has been successful in alfalfa, but only certain cultivars contain plants with resistance.

2) If the source of a pollutant such as SO_2 or HF can be pinpointed, the effluent should be reduced to an acceptable level. Usually this is a matter for public law.

Selected References

DARLEY, E. F., C. W. NICHOLS, and J. T. MIDDLETON. 1966. Identification of air pollution damage to agricultural crops. Bull. Calif. State Dep. Agric. 55:11-19.

HITCHCOCK, A. E., D. C. McCLUNE, L. H. WEINSTEIN, D. C. MacLEAN, J. S. JACOBSON, and R. H. MANDL. 1971. Effect of hydrogen fluoride fumigation on alfalfa and orchard grass: A summary of experiments from 1952 through 1965. Contrib. Boyce Thompson Inst. 24:363-386.

HOWELL, R. K., T. E. DEVINE, and C. H. HANSON. 1971. Resistance of selected alfalfa strains to ozone. Crop. Sci. 11:114-115.

JACOBSON, J. S., and A. C. HILL (eds.). 1970. Recognition of air pollution injury to vegetation: A pictorial atlas. Air Pollut. Control Assoc., Pittsburgh.

THOMPSON, C. R., G. KATES, E. L. PIPPEN, and W. H. ISOM. 1976. Effect of photochemical air pollution on two varieties of alfalfa. Environ. Sci. Technol. 10:1237-1241.

(Prepared in most part by R. K. Howell)

Herbicide Injury

Herbicide injury to alfalfa plants may be caused by excessive rates, overlapping, nonuniform application, improper incorporation, residual carry-over in the soil, drift from adjacent areas, contaminated equipment, or use of the wrong herbicide.

New Seedlings

EPTC and benefin are commonly incorporated in the soil as a preplant treatment when alfalfa is sown without a companion crop. Injury is possible if the soil is excessively sandy or low in organic matter or if the soil is dry and the temperature is high at planting time. The seedlings are stunted and the leaves may be malformed due to excessive root inhibition, but the plants usually recover and grow normally.

MCPA and 2–4D are often used at low rates to control broad-leaved weeds in fields sown with alfalfa and a companion crop. The companion crop and weeds must be dense and tall enough to form a protective canopy over the alfalfa. Typical damage from these two herbicides is shown in Plates 9 and 10.

Residual carry-over in the soil from herbicides applied to the previous crop may damage alfalfa seedlings. The excessive carry-over may be due to high application rates, overlapping, dry soil conditions, or excessively cool weather that can slow normal herbicide breakdown. Excessive carry-over from atrazine inhibits photosynthesis in the seedlings, and they die from the top down after the food reserves in the cotyledons are exhausted. In less severe cases the seedlings may recover. Injury is most likely to occur in soils low in organic content or high in pH.

Established Stands

The herbicide 2–4DB may be used for broad-leaved weed control in established stands and in seedling stands when the plants are in the trifoliolate stage. It usually causes no injury if used at the recommended rates.

Other herbicides such as simazine, pronamide, or terbacil are applied to established stands one year old or older. These herbicides must be applied when the alfalfa is dormant. Injury may appear if the timing is wrong, in sandy soils, or in soils low in organic matter or high in pH.

Herbicide drift from treating adjoining fields, fence rows, or roadways is rather common and may cause damage. The injury is more severe at the field margins adjacent to the sprayed area. Damage from drift is usually less severe than damage from direct application, and the plants usually recover.

Other Abiotic Agents and Genetic Abnormalities

Excess salt (fertilizer) accumulations may injure or kill alfalfa seedlings in the southeastern USA when an extended dry period follows emergence. Salts increase in

Figure 17. Physiogenic spotting and chlorosis of alfalfa leaf (right); healthy leaf (left). (From files of K. W. Kreitlow, USDA)

the upper few centimeters of soil through evaporation. Seedlings have a light tan, scorched appearance, especially at the leaf margins.

Minute brown to black leaf and stem spots may appear on alfalfa and other legumes grown in the glasshouse (and rarely in the field). The minute spots fail to develop on plants grown above 23°C and at low light intensity (Fig. 17). The exact cause of the spotting is unknown. Attempts to isolate a biotic causal agent have failed.

Figure 18. Vestigial flowers and branched racemes, a genetically controlled character. (Courtesy D. K. Barnes)

Various types of white spotting and irregular leaf necrosis have been described. Some of these disorders appear to be genetically controlled since they are associated with certain breeding lines or cultivars (Plate 11).

The term "vestigial flower and branched raceme" applies to plants on which the flowers have no petals and the racemes are much branched (Fig. 18). The color is greenish yellow. The abnormality, also known as carrot top, is genetically controlled and does not affect forage or seed production because the incidence of affected plants is low. Because such plants produce little or no seed, the incidence of affected plants is not likely to increase.

Selected References

AZIZI, M. R., and D. K. BARNES. 1977. Characterization and inheritance of a spotted leaf trait in alfalfa. Crop Sci. 17:126-132.

DUDLEY, J. W., and C. P. WILSIE. 1956. Inheritance of branched raceme and vestigial flower in alfalfa. Agron. J. 49:320-323.

KREITLOW, K. W., and R. A. KILPATRICK. 1967. A physiogenic leaf and stem spot of some forage legumes. Plant Dis. Rep. 51:619-622.

Part II. Diseases Caused by Biotic Agents

Bacterial Diseases

Bacterial Wilt

Bacterial wilt undoubtedly existed in the USA for many years before it was recognized as a distinct disease in Wisconsin and Illinois in 1924. The causal agent was identified and described in 1925, and within a year the disease was reported from 10 additional states. It is now known to be a problem wherever alfalfa is grown in North America, except in the desert areas of the Southwest and in areas of low rainfall without irrigation.

The bacterium was probably introduced in seed from Turkestan, and the disease now occurs in Canada, Mexico, Chile, Europe, the USSR, the Near East, Japan, Australia, and New Zealand.

Symptoms

Infected plants are usually scattered throughout a stand and are easily detected by their yellow-green color and stunted growth. Mild symptoms consist of leaf mottling with slight cupping or upward curling of the leaflets and some reduction in plant height. Severely infected plants are stunted and yellow-green, with many spindly stems and small, distorted leaflets (Plate 12A). Diseased plants are usually most evident in the regrowth after clipping.

Cross sections of taproots show a yellowish brown discoloration of the outer vascular tissue; as the disease progresses, the entire stele discolors. When the bark is peeled away, the stele is yellowish brown, in contrast to the white of healthy plants (Plate 12B). Pockets of infection sometimes appear on the inner surface of the bark (Plate 13).

Causal Organism

The disease is caused by the bacterium *Corynebacterium insidiosum* (McCull.) H. L. Jens. (syn. *Aplanobacter insidiosum* McCull.).

The bacterial cells are short rods, rounded at the ends, and 0.7–2 × 0.4–0.5 μm. Irregular shapes are common. They are Gram-positive, nonacid-fast, nonsporeforming, and capsulate in most media (Fig. 19). Growth is slow in culture. The colonies are first white, then pale yellow. Blue granules form in media containing sugar, giving the colonies a bluish color. Optimum temperature for growth on artificial media is near 23°C, minimum is below 1°C, and maximum is about 31°C. The organism grows well on beef lactose agar and produces the typical blue granules.

Disease Cycle and Epidemiology

The causal bacteria survive in plant material in the soil and usually infect plants through wounds in the roots and crown or through cut ends of newly mowed

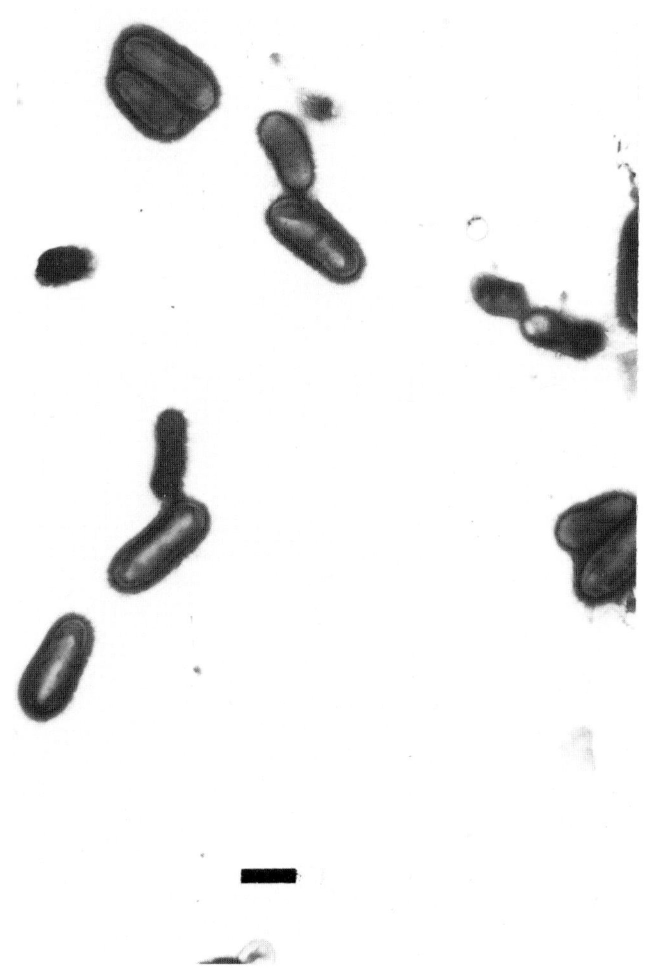

Figure 19. Electron transmission micrograph of cells of *Corynebacterium insidiosum,* the cause of bacterial wilt. Note binary fission. Bar represents 1 μm. (Courtesy L. E. B. Johnson)

stems. The wounds may be caused by winter injury, soil fauna, or mechanical injury. The bacteria enter the parenchymatous tissue and multiply intercellularly. From there, they enter the vessels and move up or down with equal ease. The symptoms and death of the plants are believed to result partly from production of a phytotoxic glycopeptide by *C. insidiosum* and plugging of the vessels. Once infection occurs, susceptible plants generally do not recover. Disease symptoms do not usually appear before the second or third crop year.

The bacteria can survive in dry plant tissue or seed for at least 10 years in the laboratory. Long-distance dissemination most likely occurs in seed and hay. The causal agent may be spread in the field by surface water, soil fauna, tillage machinery, and mower.

Control

1) Grow resistant cultivars.

2) Harvest young stands before old stands when using the same equipment. Do not mow when plants are wet.

Selected References

CHO, Y. S., R. D. WILCOXSON, and F. I. FROSHEISER. 1973. Differences in anatomy, plant-extracts, and movement of bacteria in plants of bacterial wilt resistant and susceptible varieties of alfalfa. Phytopathology 63:760-765.

CORMACK, M. W. 1961. Longevity of the bacterial wilt organism in alfalfa hay, pod debris, and seed. Phytopathology 51:260-261.

HAWN, E. J. 1971. Mode of transmission of Corynebacterium insidiosum by Ditylenchus dipsaci. Nematologica 3:420-421.

JONES, F. R., and L. McCULLOCH. 1926. A bacterial wilt and root rot of alfalfa caused by Aplanobacter insidiosum L. McC. J. Agric. Res. 33:493-521.

RIES, S. M., and G. A. STROBEL. 1972. Biological properties and pathological role of a phytotoxic glycopeptide from Corynebacterium insidiosum. Physiol. Plant Pathol. 2:133-142.

Dwarf

Dwarf on alfalfa has been reported in California and Rhode Island, USA, but it may also occur in Florida in weed hosts such as bermudagrass (*Cynodon dactylon*). Dwarf is caused by the entity that causes Pierce's disease of grape (*Vitis vinifera*) and is transmitted by "sharpshooter" leafhoppers including *Draeculocephela minerva* Ball, *Carneocephala fulgida* Nott., *Helochara delta* Oman and *Neokolla circellata* (Baker). At least 20 species of leafhoppers are implicated. The incidence of the disease in alfalfa fluctuates widely in different years. For instance, before 1950 the disease was a major cause of stand loss in California and could be found readily; at present it occurs sporadically and is mainly of clinical interest.

Symptoms

Affected plants usually are stunted. Smaller than normal leaves on affected plants are dark blue-green, resembling the color of leaves on drought-stressed plants (Plate 14). The xylem tissue of the main taproot becomes brown, especially when exposed to air, and appears darker than that of plants with bacterial wilt (Plate 15). Dwarf and bacterial wilt differ distinctly in that the inside of the bark tissue of taproots affected by dwarf disease lacks discoloration.

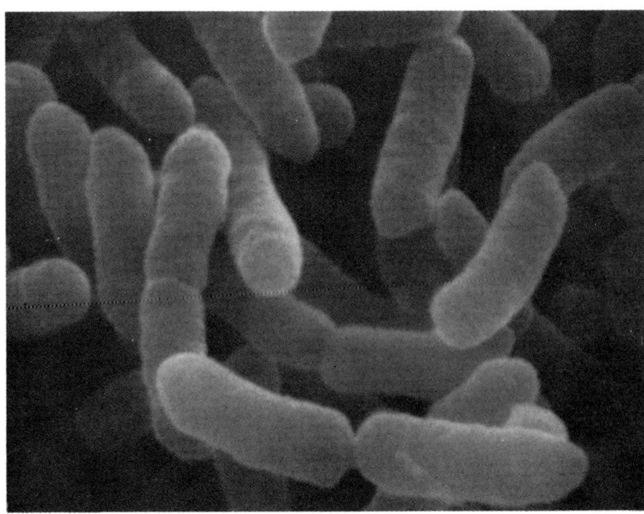

Figure 20. Unidentified fastidious bacterium that causes dwarf of alfalfa and Pierce's disease of grape, as shown by scanning electron microscopy. (Courtesy Michael Davis and Sherman Thompson)

Causal Organism

At one time, the cause was thought to be a virus because it was transmissible by grafting and by several leafhoppers. Leafhoppers transmit the causal agent from diseased alfalfa plants to grape plants, in which it causes the typical marginal leaf necrosis symptoms of Pierce's disease. Conversely, when leafhoppers transmit the causal agent of Pierce's disease from grape to alfalfa, it causes the typical dark green leaf symptoms of dwarf disease.

A rickettsialike organism with a rippled cell wall has been reported in xylem tissue of alfalfa with dwarf and grape with Pierce's disease but not in healthy plants. Recent work at the University of California, Berkeley, indicates that a Gram-negative, nutritionally fastidious bacterium causes Pierce's disease of grape (Fig. 20).

Disease Cycle and Epidemiology

Leafhopper vectors carry the causal agent to susceptible plants and inject it during feeding. The incidence of the disease varies widely. The disease is now relatively rare but was important 40 years ago in stand decline in the Riverside area and in the San Joaquin Valley of California.

Control

The cultivar California Common 49 was selected from California Common because of its resistance to dwarf. Since the reduction in incidence of dwarf disease, however, this cultivar is seldom used because it is susceptible to the spotted alfalfa aphid.

Selected References

DAVIS, M. J., A. H. PURCELL, and S. V. THOMSON. 1978. Pierce's disease of grapevine: Isolation of the causal bacterium. Science 199:75-77.

GOHEEN, A. C., G. NYLAND, and S. K. LOWE. 1973. Association of a rickettsia-like organism with Pierce's disease of grapevines and alfalfa dwarf and heat therapy of the disease in grapevines. Phytopathology 63:341-345.

HEWITT, W. B., B. R. HOUSTON, N. W. FRAZIER, and J. H.

FREITAG. 1946. Leafhopper transmission of the virus causing Pierce's disease of grape and dwarf of alfalfa. Phytopathology 36:117-128.
STONER, W. N. 1958. Field symptoms indicate occurrence of "alfalfa dwarf" or "Pierce's disease" virus in Rhode Island. Plant Dis. Rep. 42:573-580.

Bacterial Leaf Spot

Bacterial leaf spot was first reported in 1935 in Wisconsin and later identified in the eastern, central, and southwestern USA. It has been reported in widely separated countries including India, Sudan, Australia, Japan, and the USSR and probably occurs in warm areas wherever alfalfa is grown. Overall losses are minor, but locally the disease can be severe. Bacterial leaf spot is most serious in hot moist conditions, but it is often more conspicuous during hot dry periods that are unfavorable for most other foliar diseases.

Symptoms

Leaf symptoms appear first as small, diffuse chlorotic areas in which small circular water-soaked spots develop and enlarge to form irregularly shaped lesions with chlorotic margins. Water soaking is more pronounced on the lower surface of the leaves. The centers of the lesions dry to a papery texture, often shine due to dried exudate, and become light yellow to tan and translucent (Plate 16). However, fungi that invade the bacterial leaf spot lesions often alter or mask these symptoms. Severe defoliation is common.

Stem lesions begin as minute water-soaked greaselike spots that enlarge and often coalesce to form elongate lesions that may extend for several internodes. Older lesions become light to dark brown with much darker margins. In the central USA at least, lesions usually originate on the windward side of stems but rarely encircle them.

Causal Organism

Xanthomonas alfalfae (Riker et al.) Dows. (syn. *Phytomonas alfalfae* Riker et al.)

Because most xanthomonads can be readily distinguished only by their host range, some workers classify *X. alfalfae* as a *forma specialis* of *X. phaseoli* (Smith) Dows. Others include it in the *X. campestris* (Pammel) Dows. group because the xanthomonads can be distinguished from *X. campestris* only by their host reactions. The bacterium is a Gram-negative rod, 0.2–0.8 × 0.9–4.6 μm, motile by means of a polar flagellum. Colonies on nutrient agar are pale yellow, circular, smooth, and convex with an entire margin. Optimum temperature for growth is 27–32°C; maximum, above 36°C; and minimum, below 4°C. The bacterium from alfalfa will infect several other leguminous species.

Disease Cycle and Epidemiology

X. alfalfae overwinters on alfalfa debris on or in the soil and can survive for several years in hay or in debris with stored seed.

Wind and rain disperse the bacterium in the field, and it enters the plants through stomates or wounds, particularly wounds made by insects or windblown soil particles. Hot rainy weather favors bacterial leaf spot development, but the disease can become severe under hot and dry conditions. In the latter case, windblown infested soil particles probably wound and inoculate the plants.

Plants from seedings in late summer or early fall may show postemergence damping-off and severe stunting by the bacterium if high temperatures prevail.

Control

1) Resistant cultivars, being developed, offer the most practical means of control. Susceptibility among plants differs markedly within and among cultivars.

2) Spring seeding in the USA, if otherwise practical, probably would prevent seedling losses from the disease.

Selected References

BUCHANAN, R. E., and N. E. GIBBONS (eds.). 1974. Bergey's manual of determinative bacteriology, 8th ed. Williams & Wilkins Co., Baltimore, MD. 1,268 p.
CLAFLIN, L. E., D. L. STUTEVILLE, and D. V. ARMBRUST. 1973. Windblown soil in the epidemiology of bacterial leaf spot of alfalfa and common blight of bean. Phytopathology 63:1417-1419.
MOFFETT, M. L., and J. A. G. IRWIN. 1975. Bacterial leaf and stem spot (Xanthomonas alfalfae) of lucerne in Queensland. Aust. J. Exp. Agric. Anim. Husb. 15:223-226.
RIKER, A. J., F. R. JONES, and M. C. DAVIS. 1935. Bacterial leaf spot of alfalfa. J. Agric. Res. 51:177-182.
SABET, K. A. 1959. Studies in the bacterial diseases of Sudan crops. III. On the occurrence, host range and taxonomy of the bacteria causing leaf blight diseases of certain leguminous plants. Ann. Appl. Biol. 47:318-331.

Bacterial Stem Blight

Since first observed in 1904 in Colorado, bacterial stem blight has been reported in many states across the northern half of the USA, in Yugoslavia, and in the USSR. The disease is confined to the first cutting and usually is economically important in the USA only on growth damaged by a late spring freeze. The disease is sporadic, and the overall losses are minor in most areas. However, in the higher and colder valleys of western mountainous regions of the USA, first crop hay losses have been 40–50% in some cultivars.

Symptoms

Stem symptoms begin as water-soaked yellowish to olive green lesions, usually at the point of attachment of a leaf, and then extend down one side of the stem for one to three internodes. Lesions become amber-colored, glisten from dried exudate, and blacken with age. A stem may have more than one lesion, and the lower three to five internodes are most affected. The vascular tissue is not invaded. Infected stems are shorter than normal, spindly, and brittle and break easily.

Leaves become water-soaked and pale yellow at the base, along the leaflet midrib, and especially in the petiolules. Those attached to the areas of the infected stems become yellow and dry.

Causal Organism

Pseudomonas syringae van Hall, formerly referred to as *P. medicaginis* (Sackett) E. F. Smith and *Phytomonas medicaginis* (Sackett) Bergey et al.

The bacterium is a Gram-negative rod with rounded ends ($0.5–0.8 \times 1.2–2.4\ \mu m$) that is motile by one to four flagella at each end. Long chains of cells may form. Colonies on nutrient agar are smooth, circular, slightly convex, grayish white, glistening, and translucent; they produce a greenish fluorescent pigment that diffuses into the medium. Optimum temperature for growth is 27–30°C; maximum is below 37°C.

The eighth edition of Bergey's manual provisionally includes most plant pathogenic pseudomonads in either *P. syringae* or *P. cichorii* (Swingle) Stapp, which differ mainly in that *P. syringae* is oxidase-negative. The host range of *P. syringae* isolates from alfalfa is unknown, but overlapping host ranges by isolates are common in this group.

Disease Cycle and Epidemiology

Cold wet conditions in the spring favor the disease, and frost damage to spring growth is necessary for an outbreak. The bacterium lives in the soil in plant material and enters the shoots, primarily the stems, through frost cracks in the epidermis. Secondary spread may occur under favorable conditions, but in the USA the disease runs its course with the first cutting.

Control

1) Stem blight severity is related to late spring frost injury; thus, hardy frost-tolerant cultivars are affected less by the disease.

2) In severe cases, harvest the first cutting as soon as the danger of late frost is past. This removes the infested material and the regrowth will be healthy.

Selected References

ARSENIJEVIC, M., and Z. KLEMENT. 1969. Bacterial stem blight and wilt of lucerne in Yugoslavia. Acta Phytopathol. Acad. Sci. Hung. 4:111-116.

BUCHANAN, R. E., and N. E. GIBBONS (eds.). 1974. Bergey's manual of determinative bacteriology, 8th ed. Williams & Wilkens Co., Baltimore, MD. 1,268 p.

SACKETT, W. G. 1910. A bacterial disease of alfalfa. Colo. Agric. Exp. Stn. Bull. 158. 32 p.

Other Bacterial Diseases

A bacterial disease of roots of alfalfa and other legumes, caused by *Pseudomonas radiciperda* (Javoronkova) Savul., was reported from the USSR. *Pectobacterium carotovorum* (Jones) Waldee, and *Pseudomonas fluorescens* (Trevisan) Migula were recovered frcm alfalfa roots in Lithuania, USSR.

Two species of bacteria associated with discolored roots of stunted alfalfa plants were reported in eastern Pennsylvania. For the most part, symptoms seem typical of bacterial wilt caused by *C. insidiosum*. The proposed names of the two bacteria are *Pseudomonas marginalis* (Brown) Stevens var. *alfalfae* and *Erwinia amylovora* (Burrill) Winslow et al. var. *alfalfae*. The importance of these two species in alfalfa production is unknown.

Crown gall, caused by *Agrobacterium tumefaciens* (E. F. Sm. & Towns.) Conn, occurs on alfalfa in rare instances. The bacterium is a wound parasite with a wide host range. The disease could be confused with crown wart, but the galls do not contain resting bodies.

Selected References

PUIPENE, I. K. 1971. The specific composition of inducers of bacteriosis in lucerne roots sown in Lithuania. Trans. Acad. Sci. Lithuanian SSR, Series C. 2(55):75-81. (English translation)

SHINDE, P. A., and F. L. LUKEZIC. 1974. Isolation, pathogenicity and characterization of fluorescent pseudomonads associated with discolored alfalfa roots. Phytopathology 64:865-871.

SHINDE, P. A., and F. L. LUKEZIC. 1974. Characterization and serological comparisons of bacteria on the genus Erwinia associated with discolored alfalfa roots. Phytopathology 64:871-876.

Fungal Diseases

Pythium Seed Rot, Damping-Off, and Root Rot

Pythium spp., associated with many agricultural crops, are worldwide in distribution. Seed-rotting and preemergence and postemergence damping-off reduce stands of alfalfa, especially in wet areas. Fibrous roots of older plants are attacked to a limited extent. Seeding at a higher rate partially compensates for stand loss in the field. The disease can be devastating when seedlings are started in infested soil in a glasshouse and when alfalfa is seeded thickly in rows in research field plots.

Symptoms

During germination, seeds may be infected and their contents reduced to a brown gelatinous mass in the seed coat, or the radicle and cotyledons may become brown and soft after emergence from the seed coat. When infection occurs in later germination stages, the lesions are more typically limited to the hypocotyl and root, which become water-soaked and flaccid, then finally collapse, and dry up. The seedlings fall over (damp-off) or may become stunted, with small dark green cotyledons, and then die in a few days (Fig. 21). However, with a humid atmosphere, moderate temperatures, and sufficient soil moisture, a seedling with a diseased primary root may survive by producing secondary roots above the lesion. The closely related pathogen, *Phytophthora megasperma,* may produce the same effect.

Various *Pythium* spp. have been isolated from roots and rootlets of older plants; however, the extent of damage is believed to be minor.

Causal Organisms

Pythium spp. most commonly isolated from alfalfa seedlings include *P. debaryanum* Hesse, *P. ultimum*

Trow, *P. irregulare* Buis., and *P. splendens* Braun. Other species reported include *P. rostratum* Butl., *P. pulchrum* Mind., *P. myriotylum* Drechs., and *P. aphanidermatum* (Edson) Fitzp. Recent taxonomic work indicates that species complexes should be recognized because some species overlap. Cool to moderate temperatures favor all of the above *Pythium* spp. except *P. myriotylum* reported from Australia and *P. aphanidermatum* from Arizona, both of which are favored by high temperatures.

Sporangia of the cool weather species of *Pythium* are spherical and germinate either by formation of zoospores in a vesicle or directly by germ tubes. Oogonia are terminal or intercalary and are smooth except for *P. irregulare,* which is spiny. Oospores may or may not completely fill the oogonial cavity, depending on the species. There are usually one to five antheridia per oogonium, which arise from the oogonial stalk or from other hyphae. Oospores, too, may germinate by producing germ tubes that develop into mycelium or sporangia. *P. myriotylum* and *P. aphanidermatum* have lobate sporangia and may have more than six antheridia per oogonium.

Pythium can be isolated from infected tissue by plating well-washed tissue on water agar with a sterile kernel of corn on the agar surface. Differential media, which contain vancomycin and pimaracin, suppress bacteria and fungi other than Oomycetes. In culture, white fluffy mycelia of these *Pythium* spp. will grow across a petri plate within two days. In contrast, colonies of *Phytophthora megasperma* grow slower.

Figure 21. Seedlings stunted by *Pythium* sp. Healthy seedling at far right. (Courtesy Department of Plant Pathology, University of Wisconsin, Madison)

Disease Cycle and Epidemiology

Pythium spp. are soil inhabitants and persist on crop residue as oospores or as sporangia (Fig. 22). Some also may persist by pathogenic or saprophytic colonization of other crops and weeds. High soil moisture and, in most cases, low temperatures that are unfavorable for rapid seedling growth favor disease development. In addition to providing a water film for zoospore movement, high soil moisture may decrease host vigor, thus increasing the diffusion and availability of host exudates that stimulate spore germination. Optimum temperature is near 16°C for preemergence damping-off and between 24 and 28°C for postemergence damping-off.

Inocula exist as zoospores, directly germinating sporangia, oospores, and mycelia. Appressoria and infection pegs usually develop on the host. Penetration appears to be direct by mechanical pressure rather than by the production of enzymes.

In sand or soil culture under control conditions and apparently in the field, most alfalfa seedlings become immune after five days, and few plants older than two weeks become infected.

Control

1) Seed protectant fungicides are effective in the glasshouse but do not increase yields in the field.

2) Planting when the early growth of seedlings is maximum will reduce seedling loss.

Selected References

AMERICAN PHYTOPATHOLOGICAL SOCIETY. 1974. Symposium on the genus Pythium. Proc. Am. Phytopathol. Soc. 1:200-223.

BUCHHOLTZ, W. F. 1942. Influence of cultural factors on alfalfa seedling infection by Pythium debaryanum Hesse. Iowa Agric. Exp. Stn. Res. Bull. 296:568-592.

CHI, C. C., and E. W. HANSON. 1962. Interrelated effects of environment and age of alfalfa and red clover seedlings on susceptibility to Pythium debaryanum. Phytopathology 52:985-989.

LUMSDEN, R. D., W. A. AYERS, and R. L. DOW. 1975. Differential isolation of Pythium species from soil by means of selective media, temperature, and pH. Can. J. Microbiol. 21:606-612.

Spring Black Stem and Leaf Spot

Spring black stem and leaf spot occurs commonly in the USA, Canada, and Europe and has been reported in other alfalfa-growing areas. It is most destructive in temperate regions, causing reductions in yields and quality of the forage.

Symptoms

All aboveground parts of the plant may become infected, and the fungus may extend to the crown and upper root. In early spring numerous small black to dark brown spots develop on the lower leaves, petioles, and stems (Plate 17). Young shoots are often girdled and

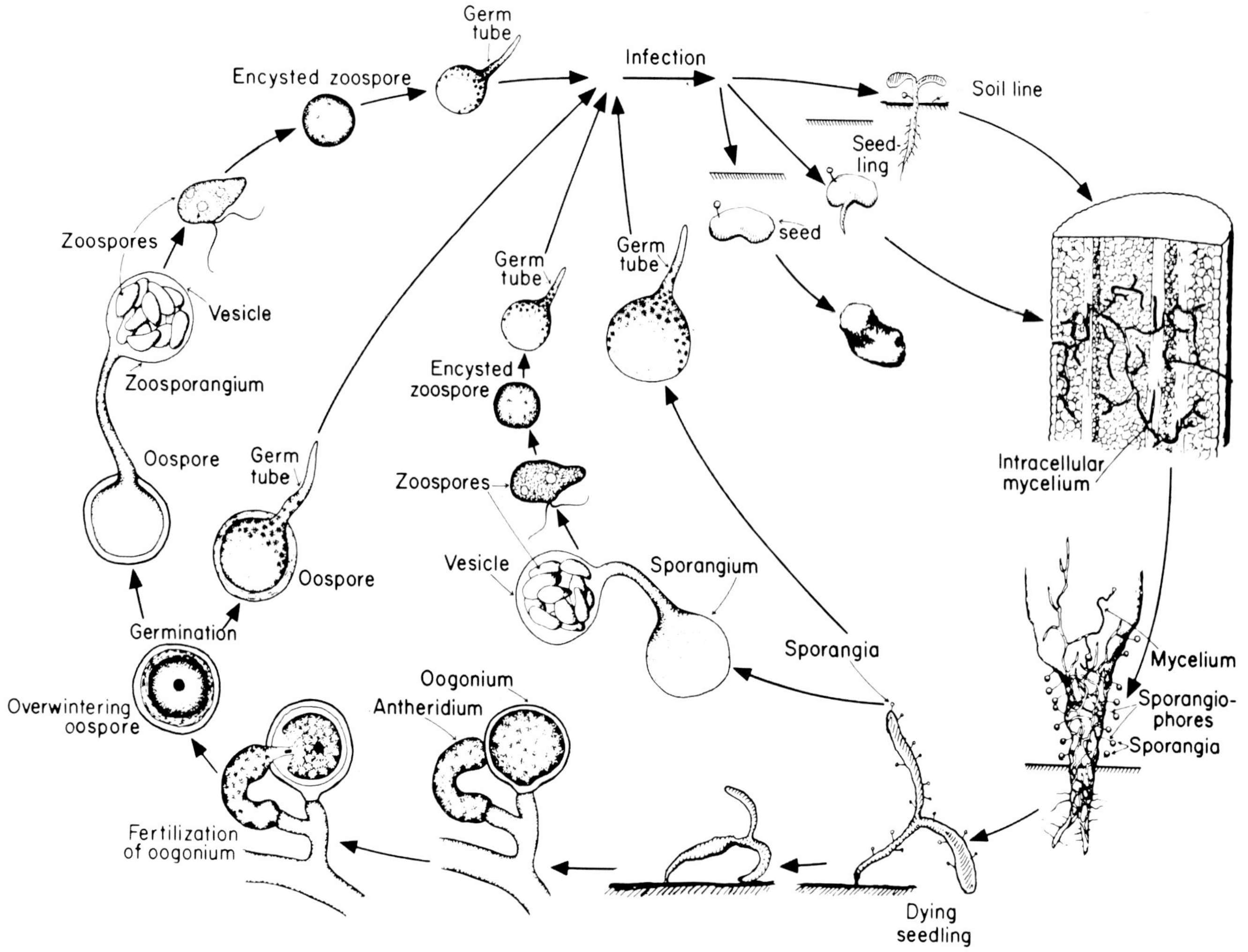

Fig. 22. *Pythium* disease cycle. (Reprinted, by permission and with modification, from Plant Pathology by G. N. Agrios. ©1969, The Academic Press, New York)

killed. Irregularly shaped to subtriangular lesions on leaves increase in size, coalesce, and become lighter brown; leaves turn yellow and often wither before falling. Lesions on stems and petioles enlarge and may girdle and blacken large areas near the base of the plant (Plate 17). The fungus also causes a crown and root rot. In humid areas seedpods may discolor and shrivel.

Causal Organism

Phoma medicaginis Malbr. & Roum. var. *medicaginis* Boerema (syn. *P. herbarum* West. var. *medicaginis* Fckl. and *Ascochyta imperfecta* Pk.)

The fungus fruits readily on overwintered stem lesions, producing brown to black pycnidia that are globose and ostiolate and range from 140 to 234 μm in diameter. When pycnidia are placed in water, the spores escape in gelatinous ropy or wedge-shaped masses. Pycnidiospores (conidia) are hyaline, oval or cylindrical with rounded ends, straight or slightly curved, continuous or uniseptate, and variable in size ($4-15 \times 2.4$ μm). A sexual stage was reported once but never verified.

On potato-dextrose agar the fungal colonies are olive green to almost black with cottony margins. Abundant pycnidia and spores are produced at 18–24°C.

The fungus forms pycnidia and pycnidiospores readily on infected leaves and stems placed in a moist chamber.

Disease Cycle and Epidemiology

Under natural conditions pycnidia rarely form on lesions during the growing season but occur in abundance on overwintered stems and fallen leaves. Involvement of the crown and upper root provides another means of carrying the inoculum through adverse conditions. Spores, which ooze from the pycnidia in the spring are spread primarily by water, but also by wind and insects. In early spring, new shoots are infected as they grow through the crop residue or stubble. Rain splashes the spores exuding from pycnidia on dead stems and spreads spores to foliage and stems. Dew or rain is necessary for pycnidia to release spores and for infection to occur. The first cutting usually is the most damaged.

In cool moist periods in the fall, the disease often builds up again. Inoculum probably comes from stubble and leaves on the ground that were killed by the fungus. In Alaska, mature pycnidia may develop in late summer.

When plants mature in the field or when seed is produced in cool humid areas, pods may become infected, and the fungus becomes seedborne as mycelium on the seed coat.

Other legumes susceptible to *P. medicaginis* under glasshouse conditions include *Medicago* spp., *Melilotus* spp., *Trifolium* spp., various peas and beans, and miscellaneous genera and species of legumes. Some workers list the fungus on clover as a separate species from that on alfalfa.

Control

1) Early cutting when spring black stem and leaf spot is prevalent tends to reduce leaf loss.

2) No highly resistant cultivars are available, but a few moderately resistant cultivars have been developed.

3) Use certified seed produced in arid areas. If seed from humid areas is used, treat it with a seed protectant.

4) Foliar fungicides effectively reduce disease severity and may benefit seed crops but are unlikely to be economical on hay crops because frequent applications are necessary to control disease. Foliar fungicides are not registered for use on alfalfa for forage in the USA.

Selected References

BOEREMA, G. H., M. M. J. DORENBOSCH, and L. LEFFRING. 1965. A comparative study of the black stem fungi on lucerne and red clover and the footrot fungus on pea. Neth. J. Plant Pathol. 71:79-89.

JOHNSON, E. M., and W. D. VALLEAU. 1933. Black stem of alfalfa, red clover, and sweet clover. Ky. Agric. Exp. Stn. Bull. 339. 82 p.

MEAD, H. W. 1964. Resume of data on black stem of alfalfa caused by Ascochyta imperfecta Peck. Can. Plant Dis. Surv. 44:134-141.

Common Leaf Spot

Common or Pseudopeziza leaf spot is one of the more destructive foliar pathogens of alfalfa and occurs worldwide. Although the disease does not kill plants, defoliation can reduce vigor, hay quality, and yield. Common leaf spot can reduce yields significantly in irrigated fields in California where more leaf loss occurs during drying and harvesting than before harvesting. In nonirrigated areas, the first or second hay crop and also the fall regrowth may be moderately damaged.

Symptoms

Small, circular, brown to black spots with dentate margins develop on the leaflets (Fig. 23). Fully developed lesions are 1–3 mm in diameter and usually do not coalesce. In the lesions are small, light brown raised fruiting bodies (apothecia) as large as 1 mm in diameter, which are visible through a hand lens. Nearly all apothecia form on the upper surface of the leaflet. Infected leaves turn yellow and drop as the disease progresses.

Causal Organism

Pseudopeziza medicaginis (Lib.) Sacc.

A single apothecium arises in a stroma beneath the epidermis. The ascus-bearing surface of the hymenial layer breaks through the epidermis, exposing the asci. Clavate asci are 50–70 × approximately 10 μm. Paraphyses among the asci are slightly longer than the asci, usually nonseptate, and often swollen at the ends. The ascospores are hyaline, ovate to oblong, nonseptate, 8–12 μm long, and approximately 5 μm wide (Fig. 24). The fungus has no known asexual stage in nature. Conidialike structures, 5–8 × 3–5 μm, are produced in culture but apparently do not play a part in the life cycle of the pathogen.

The fungus grows slowly in culture. Fructification occurs best on oatmeal agar incubated in the dark at 20°C for 21 days. The agar surface becomes covered with glistening fruiting bodies (apothecia).

Disease Cycle and Epidemiology

The leaf spot fungus overwinters in undecomposed leaves and leaf fragments on the soil surface. When the weather moderates in the spring and moisture is adequate, ascospores develop and are forcibly discharged (usually in pairs) into the air and onto alfalfa leaves, attacking lower leaves first.

Secondary ascospore inoculum comes from infected leaves on the plant and from fallen leaves. The disease incidence decreases during dry, warm summers and builds up again in autumn. Plants that are not harvested on schedule are damaged most. The pathogen is not known to be seedborne.

A closely related species of *Pseudopeziza, P. trifolii* (Biv.-Bern.) Fckl., occurs on red clover (*Trifolium pratense*), and two form species have been identified on other species of *Medicago*.

Figure 23. Common leaf spot (*Pseudopeziza medicaginis*). (Courtesy R. A. Kilpatrick, USDA)

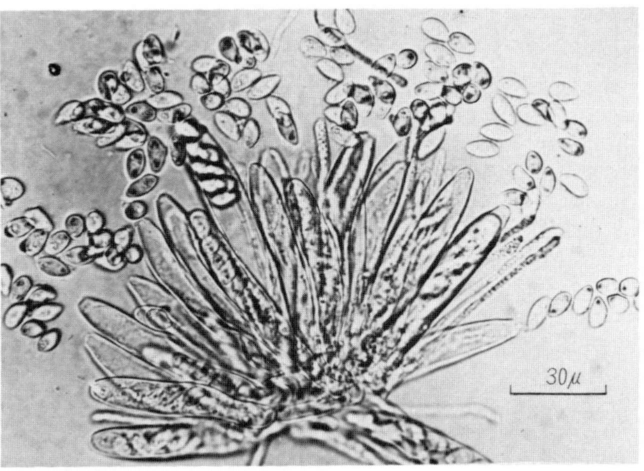

Figure 24. Asci and ascospores of *Pseudopeziza medicaginis*. (Courtesy N. Nishihara)

Control

1) Some cultivars are moderately resistant. Highly resistant breeding lines have been released.

2) Harvesting should not be delayed. If possible, harvest before leaf drop to remove inoculum and save the leaves.

3) Control with fungicidal sprays is possible and may benefit seed crops, but this probably is not economical on hay crops because frequent applications are necessary. Foliar fungicides are not registered for use on alfalfa for forage in the USA.

Selected References

HART, R. I. K., and R. C. CLOSE. 1976. Control of leaf diseases of lucerne with benomyl. Pages 42-45. Proc. 29th NZ Weed and Pest Control Conf.

JONES, F. R. 1919. The leaf-spot diseases of alfalfa and red clover caused by the fungi Pseudopeziza medicaginis and Pseudopeziza trifolii, respectively. U.S. Dep. Agric. Bull. 759. 38 p.

SCHMIEDEKNECHT, M. 1958. Untersuchungen zur specialisierung von Pseudopeziza medicaginis (Lib.) Sacc. Phytopathol. Z. 32:433-450.

SCHMIEDEKNECHT, M. 1964. Mechanik und Energetik des Sporenausstosses bei Pseudopeziza medicaginis (Lib.) Sacc. Phytopathol. Z. 51:29-40.

SUMMERS, C. G., and W. D. McCLELLAN. 1975. Effect of common leafspot on yield and quality of alfalfa in the San Joaquin Valley of California. Plant Dis. Rep. 59:504-506.

Yellow Leaf Blotch

Yellow leaf blotch occurs worldwide in temperate climates. It occurs in most states in the northern half of the USA and in Canada. In the northern Great Plains, it is often the major leaf disease on alfalfa. The way the alfalfa crop is managed and weather factors such as rainfall, dew, and wind markedly affect the development and persistence of the pathogen and development of disease. Losses are in proportion to the number of leaves that fall, which may be as high as 40% at harvest time or 80% at pod formation time. Significant losses also have been reported from the USSR, Yugoslavia, and other countries.

Symptoms

Blotches are more abundant on leaves than on petioles and stems. Their detailed characteristics are nearly identical on the various leaves of one plant but differ considerably on leaves of different plants. Symptoms on leaves (Plate 18) usually appear first as small chlorotic spots on the upper surface about 10 days after inoculation.

In a resistant plant the spots remain small and become light brown within a week. In a susceptible plant, spots enlarge to chlorotic streaks and then become yellow to orange blotches. The blotches become fan-shaped as they expand between principal veins toward the leaf edge, or they may become more or less circular. Several blotches may develop on a single leaflet. Numerous small dark pycnidia appear on the blotched areas on the upper leaf surface. The pycnidia extrude white masses of tiny spores that are not known to be germinable or infectious to the plant. Then, the central portion of the blotched area becomes dark brown to black because pseudostromata and later stroma form. The affected trifoliolate leaves may drop off the plants during rain or wind or may remain attached. In either case they curl down as they dry.

Causal Organism

Leptotrochila medicaginis (Fckl.) Schüepp (syn. *Pseudopeziza jonesii* Nannf. and *Pyrenopeziza medicaginis* Fckl.); imperfect state, *Sporonema phacidiodes* Desm.

Apothecia are black, 0.2–1.0 mm wide, round or elliptical, and individually borne on broad short stalks on either side of the leaf. These arise from the dark stromata developed in the leaf blotches. At or near maturity under moist conditions, the top part of the apothecium opens to expose a light gray to yellowish brown hymenium of asci and paraphyses. Asci are noniodine-reacting and $55-75 \times 7-10$ μm. They contain eight hyaline, ovoid, single-celled spores, $7.5-11 \times 3-6$ μm in size. Paraphyses are slightly longer than the asci. Pycnidia are subepidermal locules 100–250 μm wide and as deep as 100 μm. Conidia are hyaline, one-celled, and $5-8 \times 2$ μm.

Disease Cycle and Epidemiology

Ascospores ejected from mature apothecia are the only known infective units of the pathogen. Under favorable moisture and temperature, apothecia may form within two to three weeks on well-developed yellow-blotched leaves. In the Great Plains region of North America with its occasional heavy rains, frequent high winds, and hot dry summers, apothecia occasionally form by late fall. However, they usually do not form until early spring in leaves that are on the ground. In regions with more frequent rains, high humidity, and little wind, mature apothecia may be produced throughout the summer on leaves on the ground and on leaves of plants left for seed or delayed harvest. Mature apothecia on dried leaves may discharge ascospores after a year. Ascospores are discharged for distances to 1.7 cm; 90% of them fall as single spores. Their release is favored by relative humidities above 70% and temperatures below 25°C. On overwintered leaves, ascospores may discharge until early July, after which time the ascospore inoculum must come from a summer generation of apothecia on spring-infected leaves. Apothecia formed in the fall may discharge ascospores long before spring. Ascospores germinate at 2–31°C, optimally at 7.5–21°C. Their germ tubes penetrate the host directly within 4 hr at 20°C, 8 hr at 12°C, and 24 hr at 6°C, but they fail to penetrate at 32°C.

Control

1) Harvest early to reduce the buildup of infected leaves.

2) Cultivars with *Medicago falcata* germ plasm tend to have some resistance.

Selected References

JONES, F. R. 1918. Yellow-leafblotch of alfalfa caused by the fungus Pyrenopeziza medicaginis. J. Agric. Res. 13:307-330.

SCHÜEPP, H. 1959. Untersuchungen über Pseudopezizoideae sensu Nannfeldt. Phytopathol. Z. 36:213-269.

SEMENIUK, G., and M. D. RUMBAUGH. 1976. Reaction of some perennial and annual Medicago species and cultivars to the yellow-leafblotch disease caused by Leptotrochila medicaginis. Plant Dis. Rep. 60:596-599.

(Prepared by George Semeniuk)

Leptosphaerulina Leaf Spot

Leptosphaerulina leaf spot was a minor disease of alfalfa in the USA before 1956. During the 1956–1960 period it reached epidemic proportions in the eastern and central states. It continues to be economically important during cool wet summers in the USA, Canada, and parts of Europe and Asia.

Symptoms

This leaf spot primarily affects young leaves but also attacks petioles and other aboveground parts. Leaf symptoms vary with the plant's age, stage of growth, and environment. Lesions often start as small black spots and either remain "pepper spots," as on white clover, or enlarge to oval to round "eyespots" 1–3 mm in diameter. The lesions have light brown to tan centers and darker brown borders—often surrounded by a chlorotic area (Plate 19). When conditions favoring infection and disease development coincide with rapid regrowth, lesions appear as rather large, light tan to almost white areas that coalesce to kill the entire leaf. High light intensity increases lesion size. The dead leaflets and petioles often remain attached to the stem for a time. In older growth the young upper leaves become infected and have typical symptoms but seldom die before harvesting.

Causal Organism

Leptosphaerulina briosiana (Poll.) Graham & Luttrell (syn. *Pleosphaerulina briosiana* Poll., *Pseudoplea briosiana* (Poll.) Hoehn., *Pseudoplea medicaginis* Miles). An asexual stage has not been found.

Ascocarps, which develop readily on dead leaves or infected leaves placed in a moist chamber, are 83–152 μm in diameter. They are parenchymatous, membranous, pale brown, subepidermal bodies that open by a broad pore at the leaf surface. A few large saccate asci measuring 53–98 × 31–48 μm completely fill the ascoscarps. The ascospores are 26–46 × 11–18 μm; most are muriform with three to five (usually four) crosswalls and no, one, or two longitudinal walls (Fig. 25).

Ascocarps, asci, and ascospores in culture are slightly larger than those on host tissue. Colonies on V8 juice agar are appressed and almost black. Dispersed ascospores form many small colonies around the main colony. Mature ascospores are discharged upward to the petri dish lid within five days when incubated in light at 20–22°C. Few ascospores are produced in the absence of light.

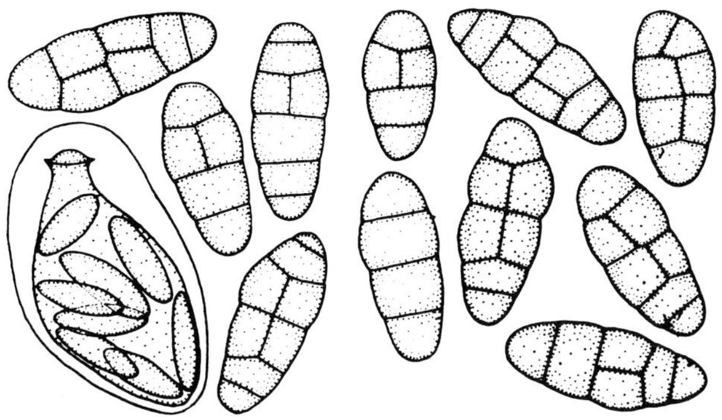

Figure 25. Ascus and ascospores of *Leptosphaerulina briosiana* from leaves of *Medicago* spp. (Courtesy E. S. Luttrell)

Disease Cycle and Epidemiology

Ascocarps develop on fallen infected leaves and provide a source of inoculum (ascospores) whenever cool moist conditions occur. Ascospores are forcibly discharged into the air currents and onto alfalfa foliage. Greatest damage occurs on leaves of young regrowth following harvest. If favorable moist weather conditions exist, however, epidemics may occur at any time during the growing season.

In northern areas the disease generally is more prevalent in the spring and early summer with less buildup again in the fall; in southern zones it is essentially a midwinter disease.

Annual species of *Medicago* also are susceptible to *L. briosiana*. Another species, *L. trifolii* (Rost.) Petr., occurs on the clovers.

Control

No effective control measures are known.

Selected References

GRAHAM, J. H., and E. S. LUTTRELL. 1961. Species of Leptosphaerulina on forage plants. Phytopathology 51:680-693.

LEATH, K. T., and R. R. HILL, Jr. 1974. Leptosphaerulina briosiana on alfalfa: Relation of lesion size on leaf age and light intensity. Phytopathology 64:233-245.

SUNDHEIM, L., and R. D. WILCOXSON. 1965. Leptosphaerulina briosiana on alfalfa: Infection and disease development, host-parasite relationships, ascospore germination and dissemination. Phytopathology 55:546-553.

Stemphylium Leaf Spot

Stemphylium leaf spot, a common foliar disease of alfalfa, may cause significant defoliation during warm wet periods in summer and fall. In the USA, the disease is more prevalent in southern and eastern states. Crop damage is difficult to evaluate because the disease is almost always associated with other foliar diseases. Foliar infection by *Stemphylium* reduces root growth and hastens death of alfalfa in the glasshouse, and some evidence indicates that similar effects occur in the field. The disease has occurred in eastern and western Europe and is widespread in New Zealand but causes little damage.

Symptoms

Spots on leaves are oval, slightly sunken, and dark brown with lighter centers; they are usually surrounded by a pale yellow halo. Older lesions may be concentrically ringed (Plate 20). Usually a single large lesion causes a leaflet to yellow and fall prematurely. The fungus can blacken stems.

In the western USA, particularly California, the pathogen produces elongate lesions with somewhat irregular outlines, tan centers, and dark brown borders (Plate 21). Lesions vary in size but seldom become longer than 3–4 mm. The disease is common during early

Figure 26. *Pleospora herbarum*. **A**, Cross section through an old leaf lesion showing conidiophores and conidia of *Stremphylium botryosum*, the asexual stage. **B**, Conidiophores and conidia. **C**, Two asci and ascospores. **D**, Germinating ascospore showing both direct and stomatal penetration. (**A** and **D** after O. F. Smith; **B** after E. G. Simmons and J. B. Ellis; and **C** after E. G. Simmons and O. F. Smith)

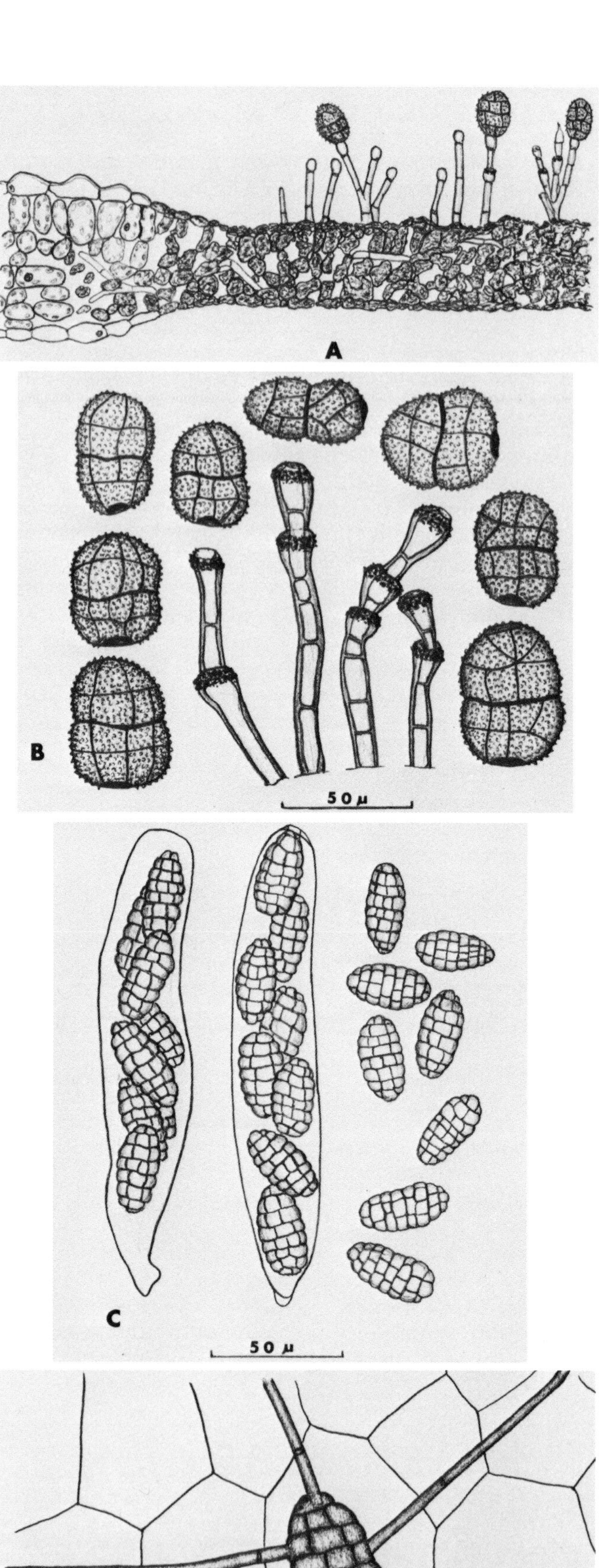

spring in interior valleys and occurs throughout the year in cooler coastal areas. Defoliation rarely results from infection by *Stemphylium*. Whether the two types of symptoms are due to biological races of the pathogen or to environmental differences is not known.

Causal Organism

Stemphylium botryosum Wallr., asexual stage; *Pleospora herbarum* (Pers. ex Fr.) Rab., sexual stage

Conidiophores and conidia develop profusely on old leaf lesions (Fig. 26A). The conidiophores are short, arise singly or in groups, and are septate and swollen at the apex. After a conidium is produced, the end of the conidiophore grows out, producing new cells and a new conidium. With this continued apical growth, a conidiophore may grow to considerable length and have a nodulose appearance (Fig. 26B). Conidia (24–40 × 14–25 μm) are olive brown, muriform, and echinulate. They are oblong with three to four septa and often constricted at the center by a median crosswall (Fig. 26B).

Perithecia are globose, membranous, and black and sometimes have a slender neck. Asci (183–267 × 27–37 μm) are oblong and cylindrical (Fig. 26C). Ascospores (32–48 × 12–21 μm) are muriform, characteristically with seven crosswalls and three to five longitudinal septa, and yellowish to brown in color (Fig. 26C).

Water induces the perithecium to rupture, after which asci are released. The asci consist of two walls. The thin outer wall bursts and the thick inner wall elongates to almost twice its original length. Ascospores are released immediately or within a few minutes.

Conidia are produced sparingly on potato dextrose agar; perithecial initials develop in 10–14 days, but ascospores are not produced until several weeks later. A temperature of 10–12°C is most favorable for ascospore production.

Disease Cycle and Epidemiology

Perithecia are difficult to find in the spring, but the ascospores they produce are undoubtedly a source of primary inoculum (Fig. 26D). The fungus also overwinters as mycelium on dead stems and leaves, because pieces placed in a moist chamber in early spring produce many conidia. The fungus is carried on seed.

Although the disease can usually be found throughout the growing season, it is most prevalent during warm moist periods in late summer and fall. The optimum temperature for conidial germination is near 25°C. The same fungus also causes a disease of red clover. Different strains may occur on other legumes and on some nonleguminous crops.

Control

1) Some cultivars have low to moderate resistance in the field.

2) Timely harvesting of forage reduces leaf loss.

Selected References

BORGES, O. L., E. H. STANFORD, and R. K. WEBSTER. 1976. The host-pathogen interaction of alfalfa and Stemphylium botryosum. Phytopathology 66:749-753.

SMITH, O. F. 1940. Stemphylium leaf spot of red clover and alfalfa. J. Agric. Res. 61:831-846.

Summer Black Stem and Leaf Spot

The disease is prevalent during hot moist periods. It is damaging to alfalfa in the USA and to some extent in South America, Africa, Asia, and Europe. During warmer than average growing seasons in the central USA, it may occur on all but the first cuttings. When the disease occurs with other stem and leaf diseases, damage from this disease is difficult to assess.

Symptoms

Small brown spots on both leaf surfaces enlarge to form roughly circular but indefinite reddish brown to smoky brown lesions 2–6 mm in diameter. When relative humidity is at or near 100% and environmental conditions favor sporulation, the lesions become ashy gray with silvery, glistening areas denoting clusters of conidiophores and conidia. Heavy infections kill leaflets and cause severe defoliation. As the season progresses, elongate dark brown stem lesions enlarge and coalesce until most of the stem is discolored. Small stems, petioles, and peduncles on nearly mature plants may die, resulting in further defoliation (Plate 22).

Causal Organism

Cercospora medicaginis Ell. & Ev.

Fascicles of 3–12 conidiophores arise from stromata on both leaf surfaces (Fig. 27A). The conidiophores are

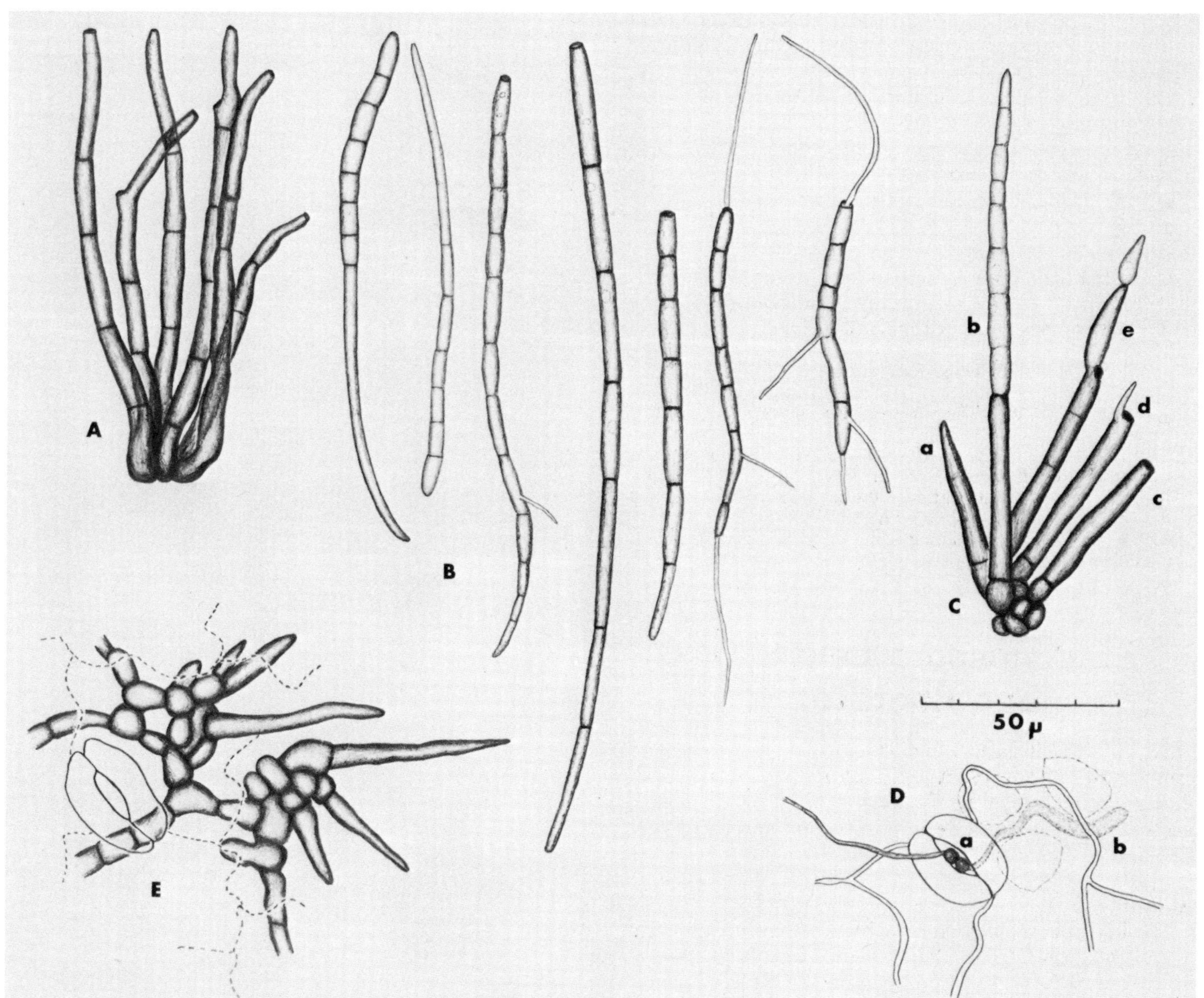

Figure 27. *Cercospora medicaginis*. **A**, Fascicle of conidiophores. **B**, Conidia showing range in length and germination. **C**, Fascicle of conidiophores showing: **a**, developing conidiophore; **b**, conidiophore with mature conidium; **c**, conidiophore with scar from a shed conidium; **d**, new conidium forming beside old scar; and **e**, conidiophore with geniculation and new conidium forming. **D: a**, germ tube in a stomate; and **b**, internal hypha with increased diameter growing under epidermis and between upper portions of palisade cells. **E**, New conidiophores from subepidermal stromata pushing through epidermis. (**A** after C. A. Schwarze, **B** after C. A. Schwarze and W. G. Willis, and **C–E** after W. G. Willis)

subhyaline to pale olivaceous brown, and septate, with no or one to four large spore scars; most are 4–5 × 15–175 μm (Fig. 27A). Conidia are hyaline, straight to slightly curved, cylindrical with the base broader than the tip, indistinctly multiseptate, and 2–4 × 40–205 μm in size (Fig. 27B). Conidiophores and conidia formed in moist chambers are much longer than those formed in drier environments. The fungus sporulates best on V8 juice agar or carrot decoction medium near 24°C.

Most workers agree that *C. medicaginis* occurs only on *Medicago* spp. and that other species attack *Trifolium* spp. (*C. zebrina* Pass.) and *Melilotus* spp. (*C. davisii* Ell. & Ev. or *Mycosphaerella davisii* Jones). However, the three *Cercospora* spp. are morphologically indistinguishable.

Disease Cycle and Epidemiology

The fungus overwinters on old infected stems and produces conidia in abundance under warm moist conditions. Severity of Cercospora leaf spot increases in proportion to the time plants are in a saturated atmosphere. Consequently, inoculum buildup increases as the plants grow taller and form a natural humidity chamber with the aid of dew. Wind and rain spread the spores to new foliage and to other plants (Fig. 27C–E). The fungus can be seedborne in warm humid areas. Secondary spread from conidia on infected stems and leaves occurs during summer and fall. The disease is most damaging when harvesting is delayed.

Control

1) Harvest the crop on schedule.

2) Resistant cultivars are not available, but some cultivars are less susceptible than others.

Selected References

BAXTER, J. W. 1956. Cercospora black stem of alfalfa. Phytopathology 46:398-400.

BERGER, R. D., and E. W. HANSON. 1963. Pathogenicity, host-parasite relationships, and morphology of some forage legume Cercosporae, and factors related to disease development. Phytopathology 53:500-508.

CHUPP, C. 1953. A monograph of the fungus genus Cercospora. Published privately by the author. Ithaca, NY. 667 p.

Stagonospora Leaf Spot and Root Rot

This disease is manifested as leaf spot, stem spot, and crown and root rot. It occurs in many areas of the USA and also in Europe. In California, the root disease is associated with stand decline in second and third year fields.

Symptoms

The leaf spot is characterized by a diffuse margin with a bleached center (Fig. 28). Infected leaves drop soon after lesions form. Pycnidia form in the infected tissue. Stem and leaf spots are similar in appearance. When large stems and taproots are cut open, the cortical and xylem tissue contains small pockets of an orange red material that gives the tissue an irregularly speckled appearance, a diagnostic symptom (Plate 23). Affected tissue is solid and dry, and even at advanced stages of root rot, little chlorosis develops on the foliage. Severely infected plants may be somewhat stunted. The fungus progresses slowly in root tissue, but often fine bright red streaks occur in the xylem well beyond the rotted areas of crown and root. In the glasshouse an incubation period of three to four months is required before symptoms appear after a single point inoculation of the crown. Externally the root becomes rough in texture and secondary invaders are common. Eventually the root becomes necrotic and the plant dies.

Causal Organism

Stagonospora meliloti (Lasch) Petr., asexual stage; *Phoma meliloti* Allesch. also has been reported. *Leptosphaeria pratensis* Sacc. & Briard, sexual stage

The *Stagonospora* pycnidia are dark brown and contain elongate spores, most of which have one or two septa (13–18 μm in length). They are often distinguished by a characteristic neck or rostrum (Fig. 29).

The ascomycetous state, *L. pratensis,* produces perithecia containing asci with septate ascospores about 28–32 μm in length.

Figure 28. Stagonospora leaf spot caused by *Stagonospora meliloti*. (Courtesy D. C. Erwin)

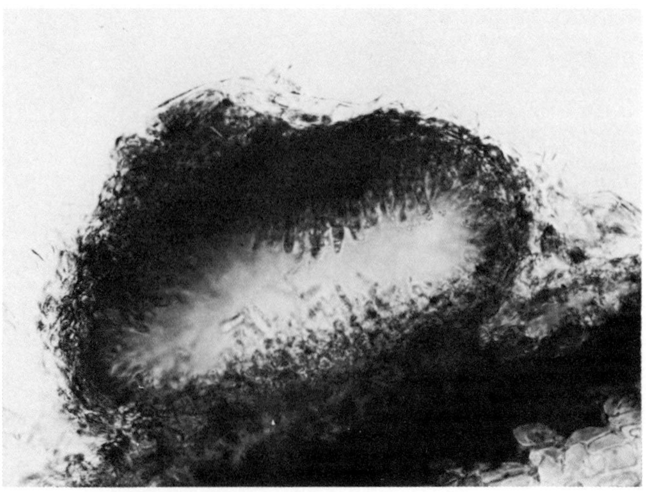

Figure 29. Cross section of pycnidium produced by *Stagonospora meliloti*. (Courtesy D. C. Erwin)

The fungus grows very slowly in culture (approximately 1 mm per day on V8 juice agar) and in host tissue. After pycnidia are formed by incubating diseased tissue on moist filter paper, the fungus can be isolated by streaking extruded pycnidiospores on V8 juice agar containing chlortetracycline (5 μg/ml). Growth of the fungus on V8 juice agar is optimum at 21–24°C, with maximum growth at 30°C. However, optimum temperature for pycnidia formation is 18–21°C. Light enhances pycnidia formation. The fungus is pathogenic only to alfalfa and to sweet clover.

Disease Cycle and Epidemiology

Little is known of the role of the *Leptosphaeria* perithecial stage, which can be found in the spring in the infection cycle. The pycnidial stage (*Stagonospora*) is readily found on lower stems and leaves, even in areas with little rainfall. The spores are extruded from wet pycnidia and are carried in irrigation or rain water to other plants. The root rot phase appears to develop from stem and crown infections and progresses slowly for two or three years in the field.

Control

Resistant cultivars are not known.

Selected References

ERWIN, D. C. 1954. Relation of Stagonospora, Rhizoctonia, and associated fungi to crown rot of alfalfa. Phytopathology 44:137-144.

JONES, F. R., and J. L. WEIMER. 1938. Stagonospora leaf spot and root rot of forage legumes. J. Agric. Res. 57:791-812.

Anthracnose

Anthracnose of alfalfa occurs sporadically and has been reported for many years in the southern half of the USA and in other countries. As early as 1947, it was observed to cause extensive damage and stand loss in alfalfa at widely separated points in the southeastern USA. More recently, probably because the use of highly susceptible Flemish alfalfas has increased, anthracnose has become more widespread and damaging. The real destructiveness was not recognized until resistant alfalfa lines were developed and compared in the field with currently grown cultivars.

In recent years anthracnose has become widespread and destructive in areas of Czechoslovakia, France, Italy, the USSR, and Australia.

Symptoms

Symptoms vary from a few irregularly shaped blackened areas on resistant stems to large sunken, oval to diamond-shaped lesions on stems of susceptible plants. The lesions are straw color with brown borders. Black fruiting structures (acervuli) develop in the bleached lesions and are readily visible with a hand lens. The lesions enlarge, coalesce, girdle, and kill one to several stems on a plant (Fig. 30). A conspicuous symptom is straw color to pearly white dead shoots scattered through the field in summer and fall.

Bluish black discoloration of invaded tissue characterizes the crown rot phase of anthracnose (Fig. 31). This symptom is often observed when killed stems are broken off at the crown. Sometimes lesions cannot be found on the dead stems. If the stem is broken off and the base is bluish black, the diagnosis is anthracnose. If the base has light brown discoloration, the disease could be Fusarium wilt or Rhizoctonia crown rot. The diseases may occur together in the same field. Other symptoms of anthracnose include blackening and killing of petioles and occasional formation of a shepherd's crook when the stem dies suddenly while wilted.

Causal Organism

Colletotrichum trifolii Bain

Scattered cushionlike masses of hyphae (acervuli) break through the epidermis in the lesion. Dark brown to black, uniseptate or continuous setae usually are visible to the eye as blackened specks against straw-colored lesions. They vary in size and number with moisture and other factors (Fig. 32). Conidiophores are hyaline, vary in length, and produce a succession of conidia apically.

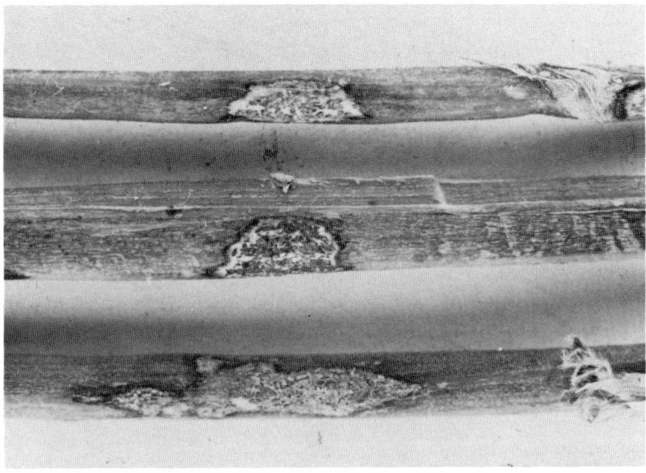

Figure 30. Anthracnose (*Colletotrichum trifolii*). Masses of acervuli are evident in stem cankers. (Courtesy D. H. Hall)

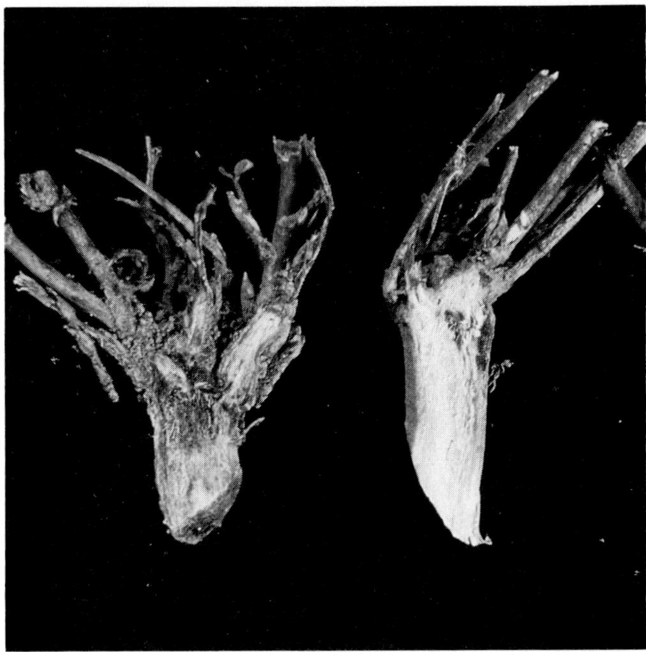

Figure 31. Crown rot caused by *Colletotrichum trifolii*. Discolored areas in the crown are bluish black. (Courtesy J. H. Graham, USDA)

Conidia are hyaline, straight, rounded at the ends, and nonseptate. The short spores are 11–15 × 3–5 μm, with a width/length ratio of 0.36–0.60 (Fig. 32). The fungus produces numerous cream to pinkish spore masses in five to seven days on V8 juice agar at 20–28°C in darkness or light.

The same fungus also causes a serious disease of red clover (southern anthracnose) in the warmer areas of the USA and other countries.

Two other species of *Colletotrichum* that cause minor damage to alfalfa are *C. destructivum* O'Gara and *C. dematium* (Pers. ex Fr.) Grove f. *truncata* (Schw.) v Arx.

Disease Cycle and Epidemiology

The pathogen can persist from one harvest season to the next in debris on the surfaces of protected harvesting equipment. This is an important means of spread from an old field to a new planting. The fungus also persists on stems and in the living stem-crown junction and in crowns of plants, probably to a greater extent in southern than in northern areas.

The fungus spreads rapidly during warm humid weather. Rain and dew wash spores from the spore masses in the acervuli onto the growing stems and petioles. The disease builds up to maximum severity between harvests in late summer and early fall. Spore masses form on stem lesions throughout the growing season and supply a ready source of secondary inoculum.

The fungus can grow down the infected stems into the crown and cause necrosis of tissue and predisposition to winter injury, or it may kill plants directly. Stands of susceptible cultivars can be reduced significantly in two or more years.

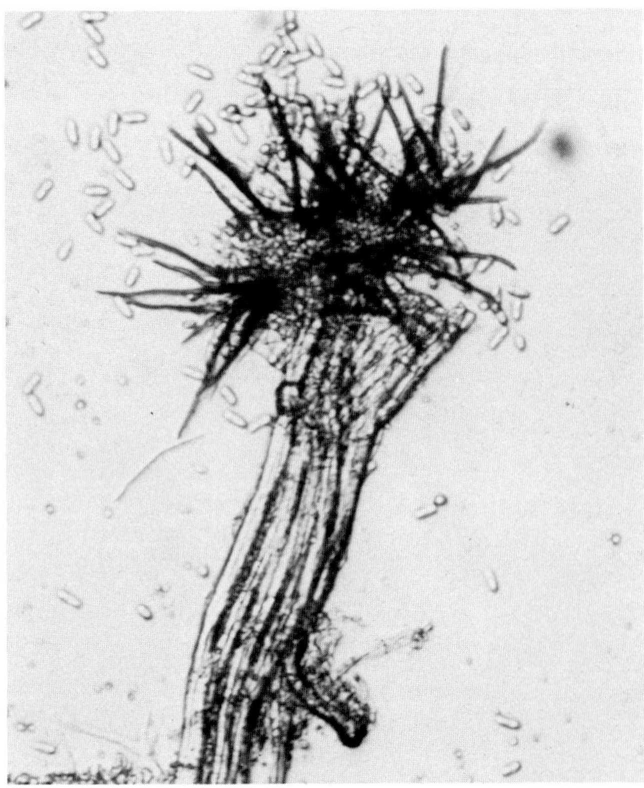

Figure 32. *Acervulus* of *Colletotrichum trifolii* that developed in crown tissue placed in a moist chamber. Note setae and conidia. (Courtesy J. H. Graham, USDA)

Control

1) Plant resistant cultivars.

2) Clean all harvesting equipment of debris before the first harvest in the spring and again during the growing season when going from fields with infected plants to other plantings.

Selected References

BARNES, D. K., S. A. OSTAZESKI, J. A. SCHILLINGER, and C. H. HANSON. 1969. Effect of anthracnose (Colletotrichum trifolii) infection on yield, stand, and vigor of alfalfa. Crop Sci. 9:344-346.

CARROLL, R. B., E. R. JONES, and R. H. SWAIN. 1977. Winter survival of Colletotrichum trifolii in Delaware. Plant Dis. Rep. 61:12-15.

GRAHAM, J. H., T. E. DEVINE, and C. H. HANSON. 1976. Occurrence and interaction of three species of Colletotrichum on alfalfa in the mid-Atlantic United States. Phytopathology 66:538-541.

IRWIN, J. A. G. 1974. Crown rot of lucerne in Queensland caused by Colletotrichum trifolii. Aust. J. Exp. Agric. Anim. Husb. 14:197-200.

LUKEZIC, F. L. 1974. Dissemination and survival of Colletotrichum trifolii under field conditions. Phytopathology 64:57-59.

Downy Mildew

This disease occurs on alfalfa in the temperate zones and at high elevations in Central and South America. Downy mildew is most severe during cool wet or humid conditions and therefore is most abundant in spring and fall. Susceptible seedlings in fall-seeded stands may be killed or so weakened that they cannot survive the first winter. Significant forage loss from mildew in established stands is usually restricted to the first cutting.

Symptoms

Diseased tissue usually becomes chlorotic. This most often includes localized areas on leaflets but may include entire leaves or shoots if infection is systemic. Invaded stems are larger in diameter and shorter in length than normal stems. Infected leaves are twisted and the margins curl down. A pale violet downy growth of conidiophores with conidia is usually conspicuous on the lower surface of infected leaves (Plate 24). Oospores are produced in the diseased tissue.

Causal Organism

Peronospora trifoliorum d By. (syn. *P. aestivalis* Syd.). *P. viciae* (Berk.) d By. has been reported on alfalfa in Mississippi.

The mycelium of *P. trifoliorum* is coenocytic, grows intercellularly in the host, and may or may not produce haustoria. Conidiophores are slender, treelike, and dichotomously branched with secondary branching. Each terminal branchlet ends as a pointed sterigma that bears a conidium (Fig. 33). Conidia are globose to broadly elliptical, 15–32 × 18–37 μm, and pale violet and germinate from the side by a germ tube. Oospores are globose, 20–37 μm in diameter, smooth or wrinkled, and light yellow. Oogonium diameter commonly exceeds oospore diameter by 10–20 μm. The fungus has not been cultured on artificial media.

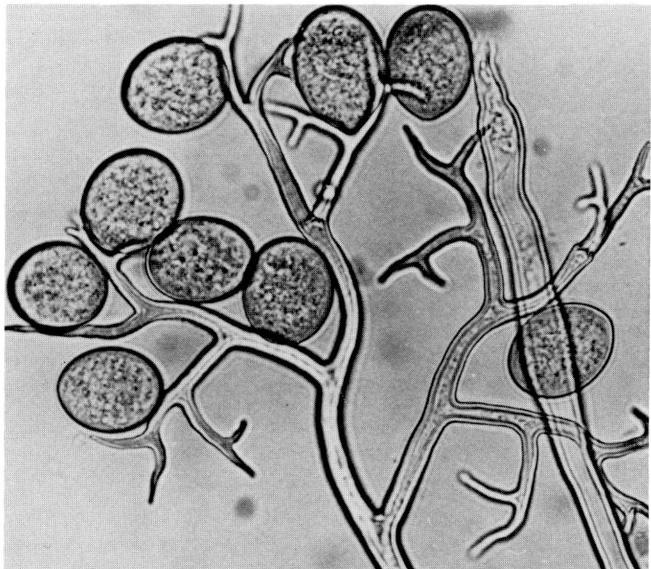

Figure 33. Conidiophores and conidia of *Peronospora trifoliorum*. (Courtesy D. L. Stuteville)

Disease Cycle and Epidemiology

P. trifoliorum overwinters in systemically infected crown buds and crown shoots of alfalfa. Oospores probably provide another means of survival, but their importance is not clear. Conidia from surviving shoots and systemically infected early spring growth provide primary inoculum. Conidia are produced only during darkness and periods of high relative humidity. They are fragile and can survive only for a few hours to a few days, depending on environmental conditions. Dissemination is primarily by wind and splashing rain. Germination occurs only in free water and at temperatures from 4 to 29°C; 18°C is optimum. A germ tube either enters through a stomate or forms an appressorium and penetrates the epidermis directly. The youngest leaves are most susceptible to infection. Secondary cycles can occur every five days under favorable conditions.

P. trifoliorum has been reported on several genera in the Leguminosae, but pathogenicity of conidia from alfalfa appears to be restricted to alfalfa and to a lesser degree to *Medicago lupulina*. Races of the fungus occur on alfalfa.

Control

1) Grow resistant cultivars. Resistance among cultivars differs markedly.

2) Spring seeding reduces the chances of seedling stand reduction from mildew in most areas.

3) Do not postpone harvest of the first cutting if mildew is severe. Harvesting removes the inoculum sources as conidia are short-lived, removes most of the young susceptible leaves, and reduces the high relative humidity provided by the plant canopy. The increase in normal seasonal temperature greatly reduces the chance of mildew in the next cutting.

Selected References

FRIED, P. M., and D. L. STUTEVILLE. 1977. Peronospora trifoliorum sporangium development and effects of humidity and light on discharge and germination. Phytopathology 67:890-894.

GÄUMANN, E. 1923. Beitrage zu einer monographie der gattung Peronospora Corda. Beitr. Kryptogamenflora Schweiz. 5:174-222.

MELHUS, I. E., and M. K. PATEL. 1929. Study of Peronospora trifoliorum DeBary on species of Leguminosae. Proc. Iowa Acad. Sci. 36:113-119.

Rust

Rust occurs on alfalfa throughout the world and sometimes damages seed production fields and early fall foliage. The disease is most widely distributed and most damaging locally in southern alfalfa-growing regions. Although rust is considered a minor disease in North America, losses do occur particularly if harvest is delayed and in some areas on seed crops. Rust has appreciably damaged crops in South Africa, the southern USSR, Israel, and the Sudan.

Symptoms

Rust is most prevalent during summer and fall. Reddish brown masses of urediospores form in pustules that rupture the epidermis of leaflets, petioles, and stems (Plate 25). When abundant, rust pustules cause the leaves to shrivel and fall prematurely. Stems are attacked severely when harvesting is delayed.

On *Euphorbia* spp., the alternate host, aecia develop on leaves and stems and may kill the stem apices. Witches' broomlike growth and occasionally hypertrophy of the stem result.

Causal Organisms

Uromyces striatus Schroet. (syn. *U. striatus* var. *medicaginis* (Pass.) Arth.)

On alfalfa, urediospores are one-celled, globoid to broadly ellipsoid, 15–28 × 14–23 μm with 1–2 μm walls, evenly echinulate, yellowish brown with three to four equatorial pores, and on hyaline pedicels (Fig. 34A). Teliospores formed later in uredia or in telia are one-celled, ovoid to broadly ellipsoid, 15–26 × 13–24 μm, and dark brown with longitudinal striations, an apical pore, and a hyaline spore stalk (Fig. 34B).

On *Euphorbia* spp. (primarily *E. cyparissias*),

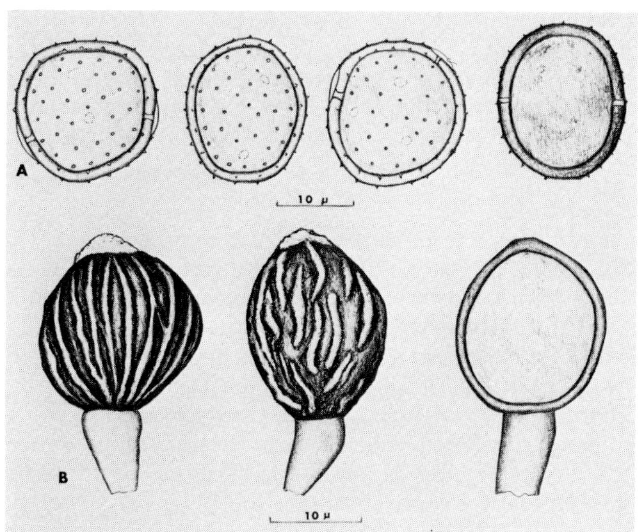

Figure 34. Rust (*Uromyces striatus*). **A**, Four urediospores. **B**, Three teliospores. (After J. C. Arthur, T. Savulescu, and J. A. Parmelee)

pycniospores are hyaline, nonseptate, ellipsoid, and 2–3 × 1–2 μm. Aeciospores are globose to broadly ellipsoid, 14–26 × 11–21 μm, and finely verrucose; they have yellow-orange contents, hyaline walls, and indistinct pores.

Races of *U. striatus* have been reported. The fungus is limited primarily to *Medicago* spp.

Disease Cycle and Epidemiology

In the southern USA urediospores or mycelium or both in plant tissue apparently persist over winter and initiate infection as the weather moderates. Urediospores spread north during the summer. Secondary spread during summer and early fall is from the spore masses (in pustules) on leaves and stems. Teliospores develop in uredia or telia, but they have no function in the absence of the alternate host (*Euphorbia* spp.).

In parts of Europe and Canada, the alternate host becomes rusted with pycniospores and aeciospores. The mycelium is perennial and systemic. The aeciospores infect nearby alfalfa earlier in the season than plants farther from the inoculum source.

Control

1) Plant resistant cultivars.
2) Harvest alfalfa on schedule to reduce leaf loss.

Selected References

DICKSON, J. G. 1956. Diseases of field crops. McGraw-Hill Book Co., Inc., New York. 517 p.

PARMELEE, J. A. 1962. Uromyces striatus Schroet. in Ontario. Can. J. Bot. 40:491-510.

Crown Wart

In alfalfa affected by crown wart, crown buds enlarge to form galls. The disease was first reported from Ecuador in the 1890s and since then has been found widely distributed in the warm areas of Europe (Germany, Italy, England, and France) and the western USA. In the USA, it is most abundant west of the Sierra Nevada and Cascade Mountains (California and Oregon), but it has been found as far east as Pennsylvania. The disease is usually confined to fields with excessive soil moisture during early spring months, and although not an important disease, it causes damage in the few fields where it occurs.

Symptoms

Galls result from infection of young crown buds (Fig. 35). Most of these originate at or slightly below the soil surface and are not noticeable. Some of the larger galls reach the surface where they take on a green color and in extreme cases form a crust of diseased tissue around the bases of stems. Ordinarily, the galls are white when young and gray to brown as they dry and decay in midsummer. The swollen cells of the gall tissue contain the resting spores of the fungus in various stages of development. When cut open, the galls are usually a mottled brown. All parts of the buds are involved—scales, leaves, and stipules. Infrequently, young leaves develop small blisterlike galls from local infections.

Causal Organism

Physoderma alfalfae (Lagh.) Karling (syn. *Urophlyctis alfalfae* (Lagh.) Magn., and *Cladochytrium alfalfae* Lagh.)

The pathogen exists in the galled tissues. The branched haustoria, with their unusually definite localization either as a solitary apparatus at the apical end of the vegetative cell or arranged in a well-defined zone between the equator and the distal pole of the resting spore, constitute the most striking morphological feature. Enlargement of the axial haustorial element initiates the resting spores from the apex of the vegetative cell. Mature resting spores are thick-walled and light brown to golden yellow.

Disease Cycle and Epidemiology

The fungus passes through summer droughts by means of the resting spores in the galls. Disintegration of the galls liberates the spores, which serve as a source

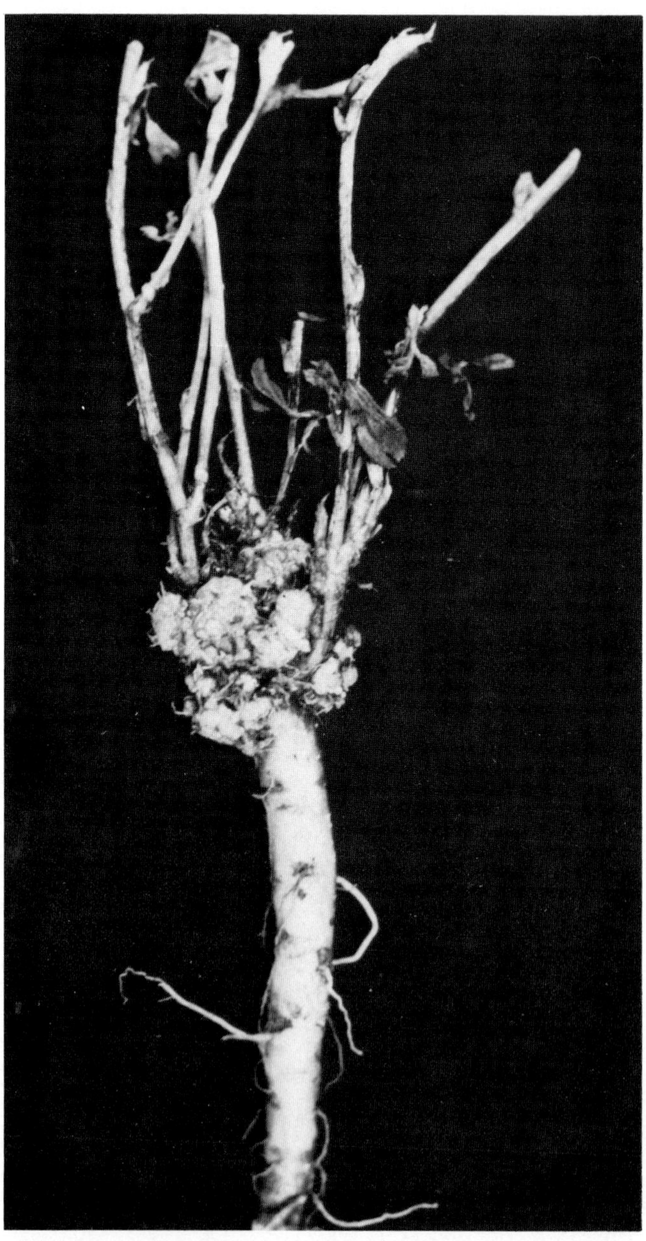

Figure 35. Crown wart (*Physoderma alfalfae*). (Courtesy D. C. Erwin)

of inoculum the following spring. The galls reach full development early in the summer. If moisture is abundant, most of the galls decay, but when dry conditions prevail in late summer, many of the galls do not decay but remain alive through the fall and winter. Penetration of the host has not been observed, but apparently the resting spores yield zoospores that germinate and thus penetrate the young buds of the host.

Control

1) Provide good drainage. Avoid excessive use of water in irrigation. Any measure that reduces moisture near the surface of the soil during the infection period should reduce the incidence of the disease.

2) Avoid following alfalfa with alfalfa on infested land.

Selected References

JONES, F. R., and C. DRECHSLER. 1920. Crownwart of alfalfa caused by Urophlyctis alfalfae. J. Agric. Res. 20:295-323.

KARLING, J. S. 1950. The genus Physoderma (Chytridiales). Lloydia 13:29-71.

LEATH, K. T. 1978. Crown wart of alfalfa in Pennsylvania. Plant Dis. Rep. 62:621-623.

Phytophthora Root Rot

Phytophthora root rot occurs in nearly every area of the world where alfalfa is grown and is especially severe in regions where alfalfa is irrigated by flooding. The disease is favored by free water in the soil resulting from irrigation or abundant rainfall, often associated with inadequate vertical or horizontal drainage. The disease can be particularly devastating to seedling stands but may also cause severe stand loss in established fields. The disease undoubtedly existed before it was first described in 1953.

Symptoms

Infected plants wilt, and the foliage, particularly the lower leaves, becomes yellow. Regrowth of diseased plants is often slow after cutting. Lesions with diffuse margins on the taproots are yellow to brown and usually start where a lateral root emerges. The yellow discoloration of tissue that extends through the root cortex into the xylem tissue is a diagnostic feature of the disease. Taproots of numerous surviving plants in the field may be rotted off at various depths, and if conditions do not favor the pathogen, new roots form (Plate 26).

Causal Organism

Phytophthora megasperma Drechs. (formerly described as *P. cryptogea* Pethyb. & Laff.)

Optimum temperatures for growth are 24–27°C; maximum is 30°C. Another *Phytophthora* isolate, *P. megasperma* cultivar HTI (high-temperature isolate), from the hot desert areas of the southwestern USA has higher optimum (27–33°C) and maximum (39°C) temperatures for growth.

P. megasperma is most readily isolated by placing small pieces of newly infected (reddish brown) tissue on water agar, dilute V8 juice agar (1:10), or on selective media. The frequency of positive isolation is often as low as 1–10%. The fungus can be detected in soil by incubating germinated alfalfa seedlings, which were pinched with a forceps above the cotyledons to slow growth, in a sample of saturated soil. After a few days, microscopic examination of the seedlings will reveal sporangia produced by the pathogen if present in the soil. Sporangia are nonpapillate and ellipsoid (Fig. 36A). These can be induced to form by flooding V8 juice agar plates containing the fungus with distilled water or salt solutions. New sporangia proliferate from within the old sporangial wall after biflagellate zoospores emerge. Sporangia also may germinate directly by formation of a germ tube to produce more mycelium or, in water, another sporangium. The fungus produces

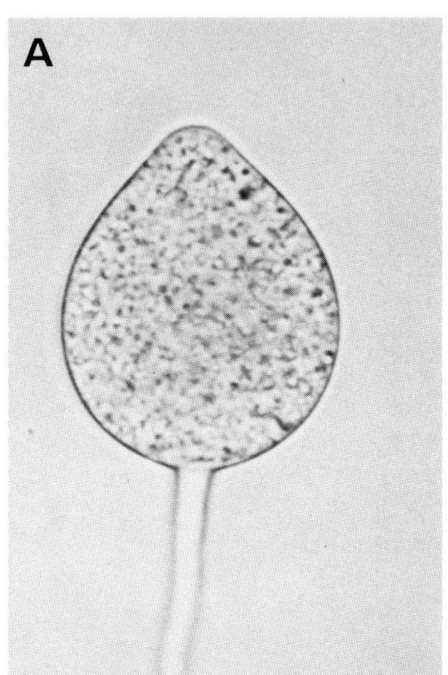

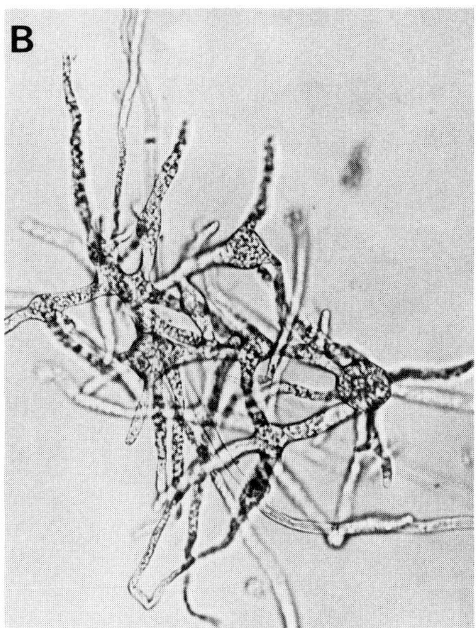

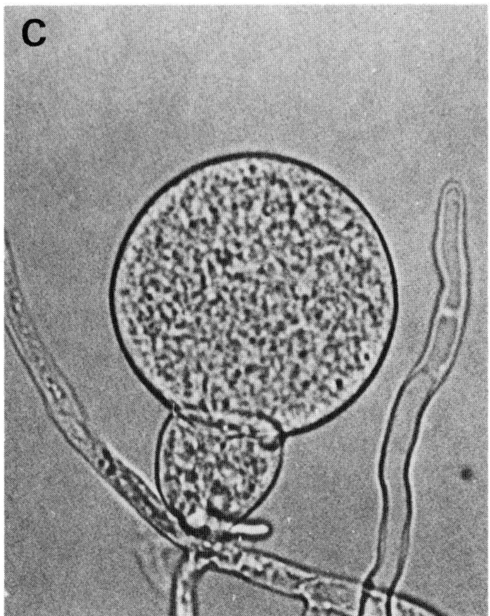

Figure 36. Morphology of *Phytophthora megasperma*. **A**, Nonpapillate sporangium. **B**, Intercalary hyphal swellings. **C**, Oogonium with amphigynous antheridium attached. The same culture also produces paragynous antheridia. A thick-walled oospore forms in the oogonium. (Courtesy D. C. Erwin)

intercalary, irregularly shaped hyphal swellings (Fig. 36B). Globose oogonia containing thick-walled oospores (about 30–32 μm in diameter) are produced homothallically in V8 juice agar or in roots of infected plants. Antheridia are amphigynous (Fig. 36C) or paragynous.

Disease Cycle and Epidemiology

Because *Phytophthora* requires free water to infect plants, Phytophthora root rot occurs only where water in the soil is excessive. The fungus probably survives between crops as oospores or as mycelium in infected plant tissue. The isolates from alfalfa are apparently host specific, except for one reported infection of garbanzo bean (*Cicer arietinum*) in a glasshouse. Optimum average soil temperatures for infection are about 24–27°C. In the Imperial Valley of California, the disease rarely occurs in summer months because soil temperatures are high. *P. megasperma* cultivar HTI was implicated there, but the disease it causes is not as severe as that caused by the moderate temperature isolates.

Irrigation water can carry zoospores that are released from sporangia formed on roots of infected plants. If waste water from affected fields is rechanneled back to the irrigation canal, the fungus can survive and be further disseminated.

Control

1) Resistant cultivars are available, and more are being developed.

2) Soil and water management are important. Deep tillage to break up compacted layers or plow soles and to improve internal percolation of water through soil will help eliminate conditions favoring infection. Land leveling, reducing the length of flood irrigation runs, and regulating the length of time water is applied will help alleviate the severity of the disease in irrigated areas.

Selected References

ERWIN, D. C. 1965. Reclassification of the causal agent of root rot of alfalfa from Phytophthora cryptogea to P. megasperma. Phytopathology 55:1139-1143.

ERWIN, D. C. 1966. Varietal reaction of alfalfa to Phytophthora megasperma and variation in virulence of the causal fungus. Phytopathology 56:653-657.

IRWIN, J. A. G. 1976. Observations on the mode of infection of lucerne roots by Phytophthora megasperma. Aust. J. Bot. 24:447-451.

MARKS, G. C., and J. E. MITCHELL. 1970. Detection, isolation, and pathogenicity of Phytophthora megasperma from soils and estimation of inoculum levels. Phytopathology 60:1687-1690.

PURSS, G. S. 1959. Root rot of lucerne. Queensl. Agric. J. 85:767-770.

RIBEIRO, O. K., D. C. ERWIN, and R. A. KHAN. 1978. A new high-temperature Phytophthora pathogenic to roots of alfalfa. Phytopathology 68:155-161.

TSAO, P. H., and S. O. GUY. 1977. Inhibition of Mortierella and Pythium in the Phytophthora-isolation medium containing hymexazol. Phytopathology 67:796-801.

Diseases Caused by Rhizoctonia

Rhizoctonia-caused diseases, which include damping-off, root canker, crown and bud rot, stem blight, and foliage blight, have been reported from all the alfalfa-growing areas of the world. Regardless of the area, most damage occurs during periods of high temperature and high soil moisture.

Both preemergence and postemergence damping-off

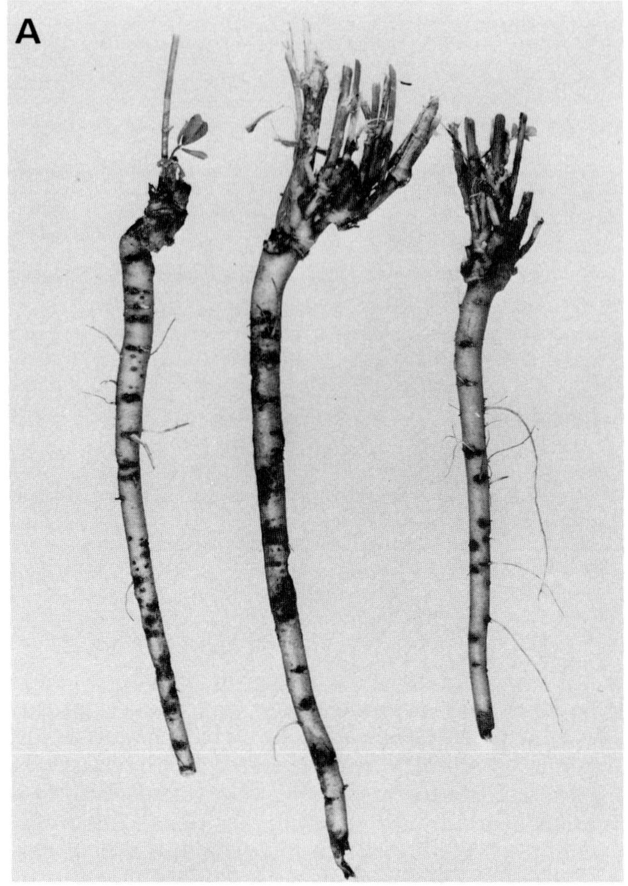

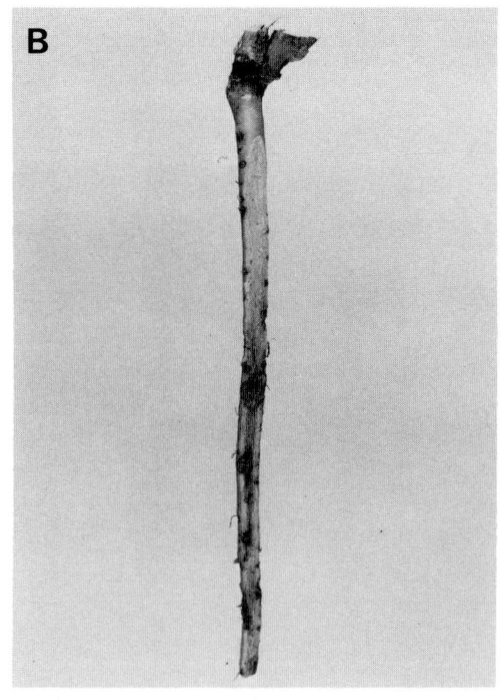

Figure 37. Rhizoctonia root canker. **A**, Elliptical to circular lesions with demarcated margins occur around areas where lateral roots emerge from the taproot. **B**, Longisection shows discoloration of xylem tissue above and below lesions. (Courtesy D. C. Erwin)

1. Nitrogen deficiency. (Courtesy Department of Agriculture, Victoria, Australia)

2. Calcium deficiency. (Courtesy Department of Agriculture, Victoria, Australia)

3. Magnesium deficiency. (Courtesy Department of Agriculture, Victoria, Australia)

4. Iron deficiency. (Courtesy R. D. Munson, Potash/Phosphate Institute)

5. Boron deficiency. **A,** Reddish leaf discoloration. (Courtesy Department of Agriculture, Victoria, Australia) **B,** Internode shortening. (Courtesy V. L. Marble)

6. Manganese deficiency. (Courtesy Department of Agriculture, Victoria, Australia)

7. Frost damage. (Courtesy J. H. Elgin, Jr., USDA)

8. Ozone injury. (Courtesy R. K. Howell, USDA)

9. Alfalfa shoot injured by the herbicide MCPA. (Courtesy F. I. Frosheiser, USDA)

10. Alfalfa shoot injured by the herbicide 2-4D. (Courtesy F. I. Frosheiser, USDA)

11. White spotting, a genetic character in some alfalfa plants. (Courtesy D. K. Barnes, USDA)

12. Bacterial wilt (*Corynebacterium insidiosum*). **A,** Healthy (left) and diseased (right) alfalfa plant. **B,** Cross sections of healthy root (bottom) and diseased roots (center and top). (Courtesy F. I. Frosheiser, USDA)

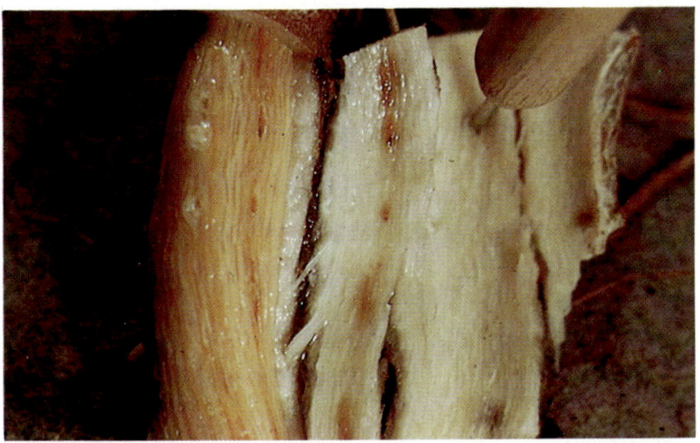

13. Bacterial wilt (*Corynebacterium insidiosum*). Discolored stele (left) and pockets of infection on inner surface of bark (right). (Courtesy D. C. Erwin)

14. Dwarf of alfalfa. Diseased plant in the foreground. (Courtesy D. C. Erwin)

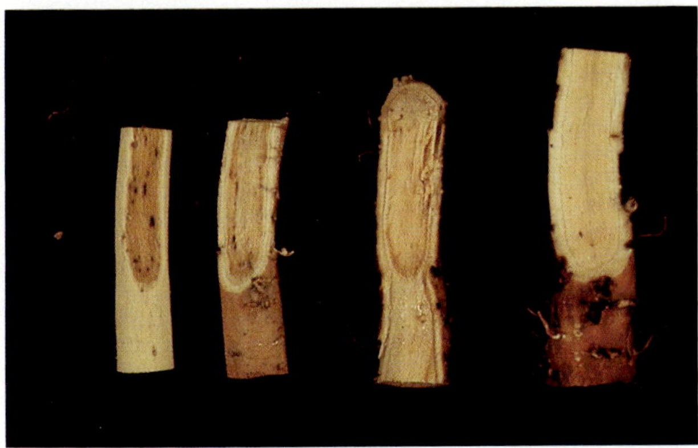

15. Tangential sections of alfalfa taproots affected by dwarf disease showing vascular discoloration. Healthy taproot (right). (Courtesy D. H. Hall)

16. Bacterial leaf spot (*Xanthomonas alfalfae*) on leaves and stems. (Courtesy R. K. Howell, USDA)

17. Spring black stem and leaf spot (*Phoma medicaginis*). (Courtesy F. I. Frosheiser, USDA)

18. Yellow leaf blotch (*Leptotrochila medicaginis*). (Courtesy D. L. Stuteville)

19. Leptosphaerulina leaf spot (*Leptosphaerulina briosiana*). (Courtesy N. Nishihara)

20. Stemphylium leaf spot (*Stemphylium botryosum*). Type of symptom produced in humid areas. (Courtesy R. A. Kilpatrick, USDA)

21. Stemphylium leaf spot from a field in California. (Courtesy D. C. Erwin)

22. Summer black stem and leaf spot (*Cercospora medicaginis*). (Courtesy D. L. Stuteville)

23. Longisection of alfalfa taproot showing typical reddish speckling caused by *Stagonospora meliloti*. (Courtesy R. A. Khan)

24. Downy mildew (*Peronospora trifoliorum*) on upper (left) and lower (right) surfaces of leaflets. (Courtesy D. L. Stuteville)

25. Uredial pustules of rust (*Uromyces striatus*) on a leaf. (Courtesy D. L. Stuteville)

26. Different degrees of severity of Phytophthora root rot of alfalfa. (Courtesy F. I. Frosheiser)

27. Stem canker caused by *Rhizoctonia solani.* (Courtesy F. I. Frosheiser)

28. Circular patch caused by the violet root rot fungus *Rhizoctonia crocorum.* (Courtesy E. W. Hanson)

29. Crown and root damaged by the violet root rot fungus *(Rhizoctonia crocorum).* (Courtesy W. G. Willis)

30. Sclerotinia crown and stem rot (*Sclerotinia trifoliorum*). Note mycelium on affected plant parts. (Courtesy F. L. Lukezic)

31. Sclerotinia crown and stem rot (*Sclerotinia trifoliorum*). Note black sclerotia. (British Crown Copyright. Reprinted by permission of Controller of Her Britannic Majesty's Stationery Office)

32. *Sclerotium* or southern blight (*Sclerotium rolfsii*). (Courtesy N. Nishihara)

33. Verticillium wilt (*Verticillium albo-atrum*). (Courtesy F. R. Sanderson)

34. Cross section of root of plant infected with *Verticillium albo-atrum.* (Courtesy F. R. Sanderson)

35. Fusarium wilt (*Fusarium oxysporum* f. sp. *medicaginis*). Longisections through taproots. Left, entire stele discolored; right, outer ring of stele discolored. (Courtesy F. I. Frosheiser, USDA)

36. Root rot of alfalfa caused by *Phymatotrichum omnivorum.* Strands of mycelium with cruciform side branches are considered diagnostic (Courtesy R. B. Hine)

37. Aerial view of circular patterns of disease caused by *Phymatotrichum omnivorum*. (Courtesy R. B. Hine)

38. Cylindrocarpon root rot (*Cylindrocarpon ehrenbergii*). Taproot is rotted for a short distance below the crown. (Courtesy F. I. Frosheiser, USDA)

39. Crown rot of alfalfa. Central core of the crown is rotted. May be caused by several fungi. (Courtesy F. I Frosheiser, USDA)

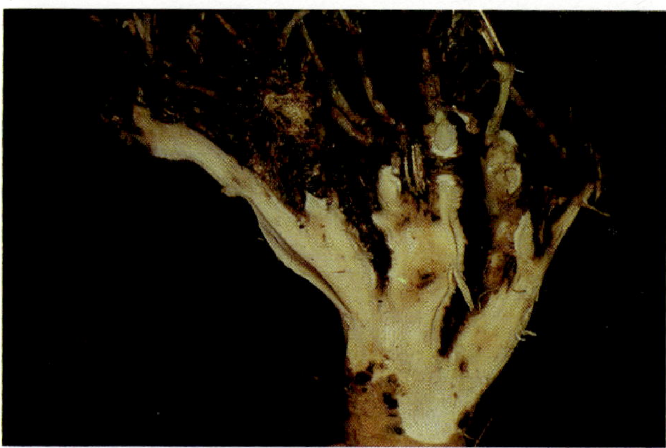

40. Brown internal lesions in alfalfa crown. An early stage of crown rot. (Courtesy D. C. Erwin)

41. Symptoms caused by alfalfa mosaic virus. (Courtesy F. I. Frosheiser)

42. Enation in an alfalfa plant affected by alfalfa enation virus. (Courtesy Bernard Alliot)

43. Witches' broom of alfalfa. Note proliferation of stems and yellow foliage. (Courtesy D. C. Erwin)

44. Discolored stem with swollen nodes and short internodes (left) caused by the alfalfa stem nematode (*Ditylenchus dipsaci*). Healthy stem (right). (Courtesy D. H. Hall)

45. White flagging caused by the alfalfa stem nematode (*Ditylenchus dipsaci*). (Courtesy G. C. Griffin, USDA)

46. Dodder (*Cuscuta* sp.). (Courtesy R. B. Hine)

47. Leafhopper burn caused by the potato leafhopper, *Empoasca fabae*. (Courtesy J. H. Graham, USDA)

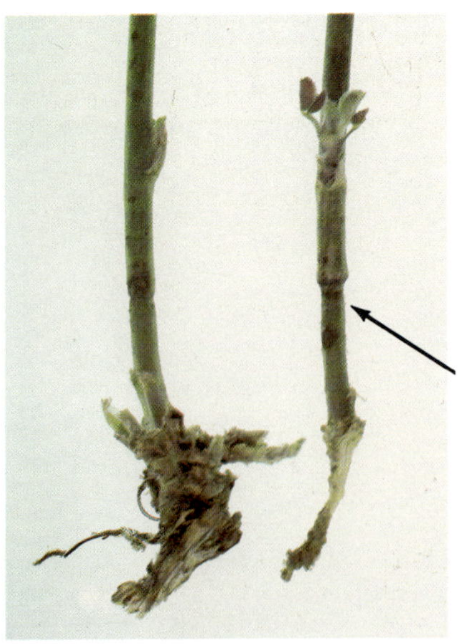

48. Stem girdle of alfalfa caused by the three-cornered alfalfa hopper. Arrow points to girdle injury. Foliage above the girdle becomes yellow to purple from blockage of phloem. (Courtesy D. C. Erwin)

occur, but the latter is more common. In the USA root canker is common in the summer in the irrigated desert regions of Arizona and California. It also occurs in Australia and in Iran under similar high temperature conditions. *Rhizoctonia solani* also is associated with crown bud rot. Crown necrosis may be due in part to *R. solani,* but a complex of other fungi such as *Fusarium* spp., *Phoma medicaginis,* and possibly others also are involved. Stem blight occurs most commonly in humid warm alfalfa-growing areas. During prolonged periods of wet hot weather, the fungus also causes a foliage blight of alfalfa.

Although some isolates of *R. solani* are favored more by high temperatures than others, the isolates that cause the various diseases do not differ morphologically.

Symptoms

In the root canker phase the fungus enters the taproot near the area where lateral roots emerge. The elliptically shaped sunken cankers are tan to buff and often are darker and more intense in color in the outer margin than in the center of the lesion (Fig. 37). During cool months of the year when the pathogen is inactive, the healed lesions blacken. The plant will die if the lesions encompass the root, but if not, new roots will emerge in the fall and support plant growth for another year.

In crown bud rot and crown rot phases, brown lesions first appear on the buds and young shoots below and at ground level. As the infection progresses, buds and shoots die and the fungus grows into the crown, preventing further production of new vegetative buds in the affected tissue (Fig. 38). Sunken, whitish to brown lesions also develop near the base of a stem and may girdle and kill it. Often brown bands develop in the older lighter lesions that indicate expansion of the infected areas (Plate 27). Coarse brown hyphae may be evident on the surface of the lesions. Rhizoctonia stem lesions can be mistaken for anthracnose but are devoid of the black discoloration (setae) common to anthracnose.

Seedling loss is most severe during high temperature and high soil moisture conditions. In contrast to *Pythium* spp. that cause damping-off of young seedlings only, *Rhizoctonia* attacks seedlings at any stage of growth. Roots and lower stems become shrunken and brown.

During prolonged periods of wet hot weather, particularly in humid areas of the southeastern USA, *R. solani* attacks leaves and stems, which collapse quickly and form a water-soaked mass of tissue. In a few days, if favorable weather conditions continue, the fungus spreads to adjacent plants. A characteristic symptom is dead leaves sticking to neighboring leaves and stems by the hyphal strands of the fungus. Sometimes large areas of dead plants appear as though killed by scalding water.

Figure 38. Crown rot caused by *Rhizoctonia solani.* (Courtesy D. C. Erwin)

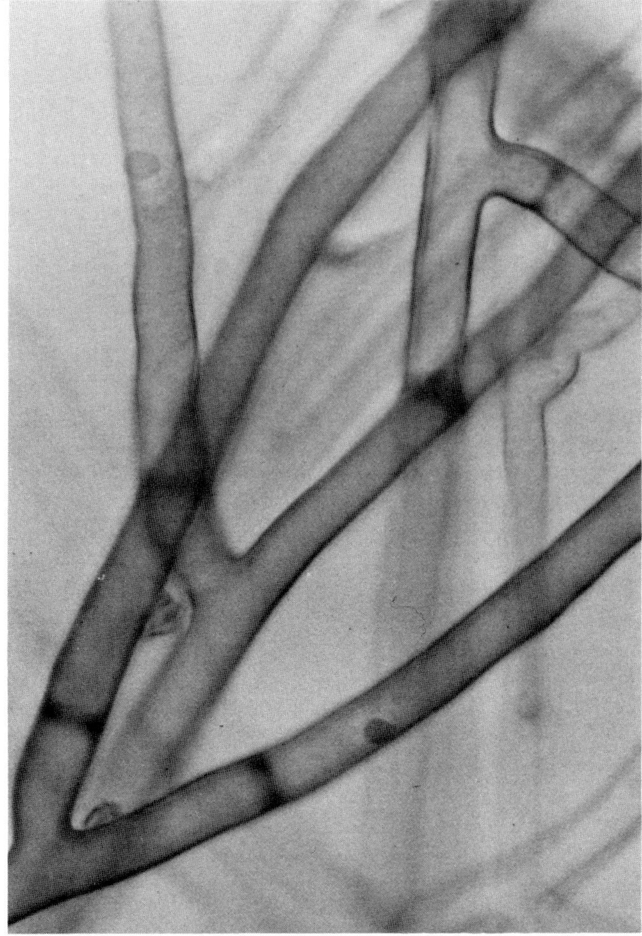

Figure 39. Photomicrograph of mycelium of *Rhizoctonia solani.* Branches from the main hyphae are slightly constricted at the base. (Courtesy D. C. Erwin)

Causal Organism

Rhizoctonia solani Kuehn. Sexual stage: *Thanatephorus cucumeris* (Frank) Donk. (syn. *Pellicularia filamentosa* (Pat.) Rogers)

The fungus grows optimally at 25–30°C, with maximum temperature at 35°C. The mycelium is brown and about 7–10 μm in diameter; it has a typical crosswall formed above the juncture of the branch. The base of the branched mycelium has a slight constriction (Fig. 39). Sclerotia are irregular in shape, less than 0.5 mm in diameter, and brown to black in color. Occasionally sclerotia can be seen on diseased plants, but this is not a dependable diagnostic feature. The basidiospore stage, which can be induced by incubating the fungus on soil in petri plates, apparently is not important in inducing the disease. Only certain strains of *R. solani* in soil can induce root canker on alfalfa; however, these are not specific to alfalfa. This characteristic is probably true also for the other diseases caused by *Rhizoctonia.*

Disease Cycle and Epidemiology

R. solani survives as sclerotia either free in soil or in plant material. The fungus is a vigorous soil inhabitant that can survive saprophytically without a host. The sclerotia germinate and the resulting hyphae form an infection cushion on the surface of the host and penetrate the plant through hyphal pegs. The fungus grows intercellularly and intracellularly and produces pectolytic enzymes in plant tissue. Plants apparently have some defense mechanisms because the round to elliptical lesions have dark delimited margins. The fungus usually penetrates young succulent tissue on seedlings or foliage and penetrates taproots through the natural wound made by emergence of a lateral root. Root canker disease occurs only in areas with high soil temperatures.

Control

1) No control measures are known. Some alfalfa selections appear to be more tolerant, probably because they are more vigorous in rooting.

2) If foliar blight develops, early cutting is suggested to reduce damage.

Selected References

ERWIN, D. C. 1956. Important diseases of alfalfa in southern California. Plant Dis. Rep. 40:380-383.

HAWN, E. J. 1958. Studies on the epidemiology of crown bud rot of alfalfa in southern Alberta. Can. J. Bot. 36:239-250.

PARMETER, J. R., Jr., H. S. WHITNEY, and W. D. PLATT. 1967. Affinities of some Rhizoctonia species that resemble mycelium of Thanatephorus cucumeris. Phytopathology 57:218-223.

SMITH, O. F. 1943. Rhizoctonia root canker of alfalfa (Medicago sativa). Phytopathology 33:1081-1085.

SMITH, O. F. 1946. Effect of soil temperature on the development of Rhizoctonia root canker of alfalfa. Phytopathology 36:638-642.

Violet Root Rot

Violet root rot is widely distributed in Europe and in North America but is rarely of much economic importance on alfalfa. It was first reported in France on *Crocus sativus* in 1728 and has since been found on many plant species; Dugger lists 54 species in 21 families. The disease was reported on alfalfa in France in 1815, in the USA in 1890, and later in Japan and other countries.

Symptoms

The disease is primarily a root rot. The mycelium of the pathogen completely invades the main root system. Aboveground parts of infected plants turn yellow and then brown and finally die. This occurs first in small circular to irregular patches that enlarge as the disease progresses (Plate 28). All or nearly all of the alfalfa plants in the interior of a patch die. Plants at the periphery of the patch are conspicuously brown, in sharp contrast to the surrounding green healthy plants.

The diseased roots have a compact felt cloak of mycelium that is bright violet to cinnamon color (Plate 29). This very diagnostic mycelium often extends 20 cm or more below the soil line and also is abundant in the crown. As the root rots, its bark loosens and its central cylinder becomes soft and shredded. Small aggregations of mycelium, the infection cushions, appear at the points of penetration by the fungus. Minute black sclerotia that are just visible to the naked eye also are often present on dying roots. The disease is most common after midsummer.

Causal Organism

Rhizoctonia crocorum DC. ex Fr. (syn. *R. violaceae* Tul.), asexual stage; *Helicobasidium purpureum* Pat., sexual stage

Mycelia of the two stages are identical. At first the mycelium is pale buff to violaceous, but it is red violet to violet brown or cinnamon when the root is completely invaded and is always violet brown with age or when densely matted. The 0.2–1.1 mm stromatic bodies distributed among the hyphae are infection cushions involved in host penetration. The true sclerotia are flattened or rounded bodies varying in diameter from a few millimeters to several centimeters. When mature, they are deep violet brown and thickly covered with a persistent velvety felt, externally of the same color as the root-invading hyphae but darker internally.

The sexual (basidial) stage apparently occurs only in the spring and requires shade and a humid atmosphere; it is extremely sensitive to high temperature, sunlight, and low humidity. The fructification consists of a fairly thick, dense, purplish or violet felt that is not gelatinous or waxy. The hymenium consists mainly, if not entirely, of basidia. The young basidium is a cylindrical erect branch arising from one of the subhymenial hyphae. Basidia are hyaline. The elongating apex of the basidium gradually bends in a characteristic crosierlike manner and soon transverse septa appear. The sterigmata arise one from each cell of the basidium. Basidiospores are hyaline and 10–15 × 6 μm in size; they may be ovate or elliptically oblong but usually are somewhat reniform.

Disease Cycle and Epidemiology

The optimum conditions for disease development are not fully understood. The disease usually appears at about the time, or soon after, the second cutting of forage is ready to be removed. The optimum temperature for mycelial growth of the *Rhizoctonia* stage and of the *Helicobasidium* stage is about 25°C.

Some researchers believe that infection occurs more

readily in sandy, well-aerated soils (provided some moisture is present) than in heavy soils, but the disease also occurs in loam soils. Many discussions of the disease mention soil acidity, but liming does not provide complete control. The disease reportedly develops rapidly under moist conditions in England and France, but it also occurs in the semiarid regions of the western USA.

The fungus may persist in the soil for three to 20 years. Young hyphae force apart the epidermal cells and infect roots; the hyphae traverse the cortex and finally reach the cambium. Eventually the taproots rot away. Infected plants left in the ground for a period of years aid the fungus in survival.

Control

1) Crop rotation is the most effect control. Susceptible crops such as alfalfa, the clovers, and birdsfoot trefoil should be spaced as widely as possible in the rotation. Corn (*Zea mays*), small grains, and forage grasses are highly resistant and can be grown safely where the disease is a problem.

2) Sanitation and good drainage have been suggested as additional control measures.

Selected References

BUDDIN, W., and E. M. WAKEFIELD. 1927. Studies on Rhizoctonia crocorum (Pers.) DC. and Helicobasidium purpureum (Tul.) Pat. Trans. Br. Mycol. Soc. 12:116-140.

BUDDIN, W., and E. M. WAKEFIELD. 1929. Further notes on the connection between Rhizoctonia crocorum and Helicobasidium purpureum. Trans. Br. Mycol. Soc. 14:97-99.

DUGGER, B. M. 1915. Rhizoctonia crocorum (Pers.) DC. and R. solani Kühn (Corticium vagum B. and C.), with notes on other species. Ann. Mo. Bot. Gard. 2:403-458.

FARIS, J. A. 1921. Violet root rot (Rhizoctonia crocorum DC.) in the United States. Phytopathology 11:412-423.

HANSON, E. W. 1963. Violet root rot of alfalfa in Wisconsin. Plant Dis. Rep. 47:306-307.

(Prepared by E. W. Hanson)

Sclerotinia Crown and Stem Rot

Crown and stem rot diseases of legumes caused by species of *Sclerotinia* were reported from Germany as early as 1857. Although a full description of the fungus, a partial host range, and some control data were published in 1872, alfalfa was not included as a host until 1915. The severity of the disease varies from season to season, and diseased plants are often scattered within plantings. Losses may involve entire fields or areas as small as 6–8 cm in diameter. Plants of all ages are susceptible, but the incidence and severity of the disease seem to be greatest in seedlings.

Disease losses occur mainly in cool humid areas including Great Britain, Germany, Norway, Sweden, Canada, and the USA. The crown and stem rot pathogen is widely distributed on forage legumes, but damage usually is less severe on alfalfa than on *Trifolium* spp., especially crimson and red clovers.

Symptoms

Infected leaves and stems become yellow and flaccid and finally collapse. A white fluffy mass of mycelium grows over the dead plant parts or the soil surface and initiates new infections (Plate 30). When the food supply is exhausted or environmental conditions are unsuitable for continued growth, the fungus produces hard black sclerotia on or in stem and crown tissue (Plate 31).

Causal Organism

The binomial proposed by Eriksson, *Sclerotinia trifoliorum* Eriks., is widely used. F. J. Seaver does not include this species in his book, "The North American Cup-Fungi (Inoperculates)," but the fungus is logically placed in the Pezizales order, Inoperculate section, Sclerotiniaceae family, and *Sclerotinia* genus, for which Seaver includes 17 species. Names for the fungus include *Peziza ciborioides* Hoffm., *S. libertiana* Fckl., *S. ciborioides* Rehm, and *Whetzelinia sclerotiorum* (Lib.) Korf & Dumont. According to L. H. Purdy's concept of the *Sclerotinia* spp., the crown and stem rot pathogen is *S. sclerotiorum* (Lib.) d By. (syn. *S. trifoliorum* Eriks.).

Sclerotia have round to irregular shapes, are black, and may be as large as 8–10 mm in diameter. Light brown apothecia are 3–8 mm in diameter with 4–25 mm stalks. The cup is depressed when young and plano-convex when mature (Fig. 40). Eight hyaline, elliptical to ovate one-celled ascospores, about $10 \times 18\ \mu m$, form in inoperculate asci separated by slender paraphyses. Nonfunctional sporelike "microconidia" form on vegetative mycelium in culture. The role of these structures is unclear, as they apparently do not germinate.

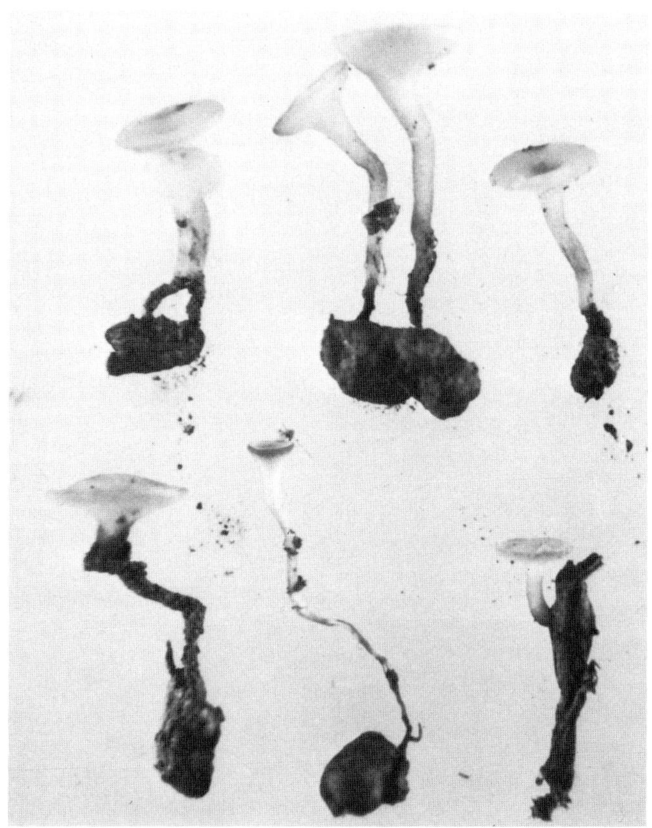

Figure 40. Sclerotia and apothecia of *Sclerotinia trifoliorum*. (Courtesy R. E. Welty)

Disease Cycle and Epidemiology

The sclerotia fall to the soil surface when the tissue collapses and may be covered with soil or plant debris. It is generally accepted that mycelium grows only to a limited extent in soil and that mycelium from sclerotia rarely initiates new plant infections. Sclerotia formed in the spring lie dormant in soil or on surface debris during the summer. In the fall sclerotia produce one to several apothecia (Fig. 40). Ascospores are forcibly discharged and are carried to leaves and stems by wind. Primary infection occurs during cool wet weather when ascospores germinate and penetrate the host directly. If high moisture and cool temperatures prevail during the winter and spring, secondary infection occurs as mycelium spreads to other leaves and stems.

The effect of temperature on growth and pathogenicity of *S. trifoliorum* is well documented. The optimum temperature for growth in culture is 15–16°C; the fungus grows between −2 and 27°C and dies at −24 and 42°C. Free water between the inoculum and the host is necessary for *S. sclerotiorum* to successfully infect bean. Further development of the disease stops if the inoculated tissue becomes dry. The same is most likely true for alfalfa.

The host range is limited largely to forage legumes and there appears to be little host specificity for isolates. In field studies of alfalfa, red clover (*Trifolium pratense*), and crown vetch (*Coronilla varia*), isolates of *S. trifoliorum* can infect alfalfa, red clover, ladino clover (*T. repens*), and crown vetch. Common hosts for *S. sclerotiorum* do not appear to be natural hosts for *S. trifoliorum;* however, seedlings of lettuce (*Lactuca sativa*), tomato (*Lycopersicon esculentum*), snap beans (*Phaseolus vulgaris*), and soybeans (*Glycine max*) can be infected in the glasshouse when favorable conditions are provided for disease development.

Infection of several hosts including red and ladino clover by ascospores of *S. sclerotiorum* is well documented. A carbon source and mechanical contact are essential for formation of appressoria, and penetration of leaves occurs in 36 hr. The mycelium lines up perpendicular to the epidermal cell and produces an infection cushion. The cuticle disintegrates, and the epidermal and parenchymal cells degenerate. After destroying the cuticle and epidermis, the fungus invades other tissues, growing intracellularly and intercellularly. Mycelium is usually restricted to the parenchymal tissues but occasionally grows in tracheids and vessels of the xylem. The fungus infects ladino clover through stomates.

Control

1) Plow deep to bury sclerotia.

2) Plant seed free of sclerotia.

3) Maintain three to four year intervals between forage legumes in rotations.

4) No alfalfa cultivars are known to be resistant to *S. trifoliorum*.

Selected References

ABAWI, G. S., and R. G. GROGAN. 1975. Source of primary inoculum and effects of temperature and moisture on infection of beans by Whetzelinia sclerotiorum. Phytopathology 65:300-309.

CARROLL, R. F., F. L. LUKEZIC, and J. M. SKELLY. 1970. Evidence that isolates of Sclerotinia trifoliorum from crownvetch and alfalfa are not specific for either host. Plant Dis. Rep. 54:811-814.

GILBERT, A. H., and C. W. BENNETT. 1917. Sclerotinia trifoliorum, the cause of stem rot of clovers and alfalfa. Phytopathology 7:432-442.

KREITLOW, K. W., and V. G. SPRAGUE. 1951. Effect of temperature on growth and pathogenicity of Sclerotinia trifoliorum. Phytopathology 41:752-757.

PRIOR, G. D., and J. H. OWEN. 1964. Pathological anatomy of Sclerotinia trifoliorum on clover and alfalfa. Phytopathology 54:784-787.

PURDY, L. H. 1955. A broader concept of the species Sclerotinia sclerotiorum based on variability. Phytopathology 45:421-427.

PURDY, L. H. 1958. Some factors affecting penetration and infection by Sclerotinia sclerotiorum. Phytopathology 48:605-609.

(Prepared by R. E. Welty)

Sclerotium or Southern Blight

Sclerotium rolfsii generally develops on a large number of hosts in tropical and subtropical countries and in the warmer areas of temperate zones when high temperature and high moisture levels prevail. In general, the disease is considered to be of minor importance in alfalfa because it is not as susceptible as many crops, but loss is occasionally serious in isolated areas on silty and sandy soils in the southern USA.

Symptoms

The fungus produces a white cottony mycelial growth over the surface of bases of stems and crown branches of alfalfa. Plant tissues above the affected area bleach to a tan color and die. Small light-brown spherical sclerotia that look like mustard seeds develop on the stems and crown and on mycelial mats that form on debris on the soil surface (Plate 32).

Causal Organism

Sclerotium rolfsii Sacc. Sexual stage: *Pellicularia rolfsii* (Curzi) West (syn. *Corticium rolfsii* Curzi)

In culture, silky white mycelia rapidly cover the agar surface. Sclerotia, about 2 mm in diameter, are produced after about a week. Maximum mycelial growth occurs between 25 and 35°C. Asexual spores have not been observed.

The sexual stage does not appear to be an important source of infection. Basidiospores formed on onion agar are elliptical to obovate, hyaline, and about $3-7 \times 6-12$ μm.

Disease Cycle and Epidemiology

Sclerotia are the principal means of overwintering. Wind, water, and infested plant debris scatter sclerotia and viable mycelia. After introduction into noninfested soil, survival of the pathogen is favored by its ubiquitous host range and rapid growth on any plant debris.

Because little work has been done on southern or sclerotial blight of alfalfa, much of this information derives from research on other crop hosts.

Selected References

AYCOCK, R. 1966. Stem rot and other diseases caused by Sclerotium rolfsii. N.C. Agric. Exp. Stn. Tech. Bull. 174. 202 p.

JOHNSON, H. W. 1968. Alfalfa diseases. Miss. Agric. Exp. Stn. Bull. 767. 12 p.

Verticillium Wilt

Verticillium wilt of alfalfa was reported in Sweden in 1918 and in Germany in 1938. Since then, the disease has spread through the alfalfa-growing areas of Great Britain and continental Europe. It is probably the most damaging alfalfa disease in northern Europe. Verticillium wilt is widespread in New Zealand, was once reported in eastern Canada, and now also occurs in Washington, Oregon, and Idaho, USA.

In European areas where the disease is prevalent, more than 50% of the plants in a field are often infected the first year. Frequently by the end of the second year and invariably at the end of the third year, the crop is nonproductive.

Symptoms

Temporary flagging of upper leaves occurs on warm summer days. The lower leaves and shoots later wilt and become pale, yellow, and ultimately bleached and desiccated (Plate 33). Normal new shoots may develop from the crown, but they wilt rapidly and die. Conidiophores may cover the base of infected stems, giving them a gray appearance. When the stem dies, the infected area may blacken.

The disease produces no external symptoms on the roots. Internally, the xylem tissue becomes light to dark brown (Plate 34). In completely wilted plants, the discoloration can often be traced from small lateral rootlets through the taproot into the stems, the petioles, and frequently the pedicles of the flower.

In the advanced stage of the disease, plants in the field are stunted with yellow and desiccated shoots and leaves. Stems often remain green after the leaves are killed.

Causal Organisms

Verticillium albo-atrum Reinke & Berth. and *V. dahliae* Kleb.

V. albo-atrum (dark mycelial species) is the principal cause of Verticillium wilt. Although *V. dahliae* (microsclerotial species) causes similar symptoms, it is less virulent and is found only occasionally in the field in England.

Conidiophores of *V. albo-atrum,* usually with swollen but rarely with darkened bases, branch verticillately (in a whorl), and conidia are borne singly at the apex of the branches (Fig. 41). Conidia are mainly $3.5–8 \times 2.3$ μm, hyaline, and nonseptate (rarely uniseptate). Dark thickened resting mycelium and dark swollen hyphae are produced on the host and in culture (Fig. 42). Optimum temperature for growth is 20–22°C; the maximum is 30°C.

To correctly identify *V. albo-atrum,* it *must* be grown in culture. Other species of *Verticillium* that are mild pathogens on crops other than lucerne (e.g., *V. nigrescens* Pethyb., *V. nubilium* Pethyb., and *V. tricorpus* Isaac) or are saprophytes (e.g., *V. lateritium* Berk.) and another fungus (*Gliocladium roseum* Bainier) that is frequently confused with *Verticillium,* all produce verticillate conidiophores and single-celled conidia.

Disease Cycle and Epidemiology

The causal agent (*V. albo-atrum*) persists over winter as dark thick-walled mycelium in plant debris, in infected plants, and possibly on weed hosts. The fungus enters the roots and inhabits the water-conducting system. Conidia are produced under cool moist conditions and are spread primarily by wind currents and in windblown debris. The fungus is carried on trash with seed and can persist only for short periods of time directly on the seed. Secondary spread during the growing season is mainly by mowing and windblown conidia and debris.

In inoculation studies, isolates of *V. albo-atrum* from alfalfa will infect a number of common weeds and remain pathogenic to alfalfa when reisolated. *V. albo-atrum* also attacks other crops to some degree, e.g., potatoes and runner beans, but isolates from these and other crop plants do not produce the typical wilt symptoms on alfalfa. Thus, *V. albo-atrum* on alfalfa is apparently a different strain from isolates obtained from other crop plants.

Control

1) Resistant cultivars have been developed in Europe. Most American cultivars are susceptible, but some may have partial resistance. Resistant cultivars are being developed.

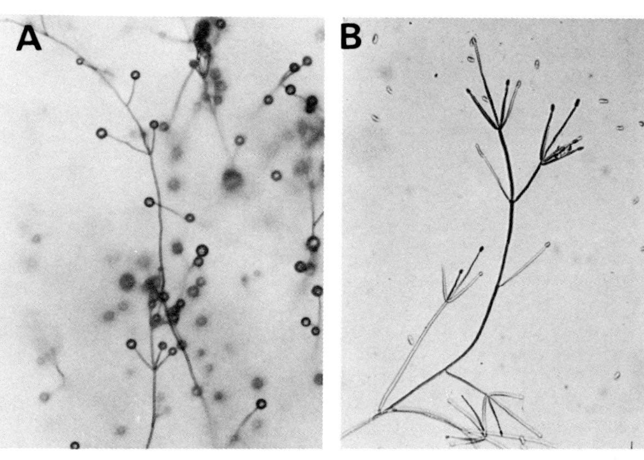

Figure 41. Conidiophores and conidia of *Verticillium albo-atrum*. **A**, Conidia in drops of water. **B**, Conidia borne singly at tips of conidiophores. (Courtesy F. A. Uecker, USDA)

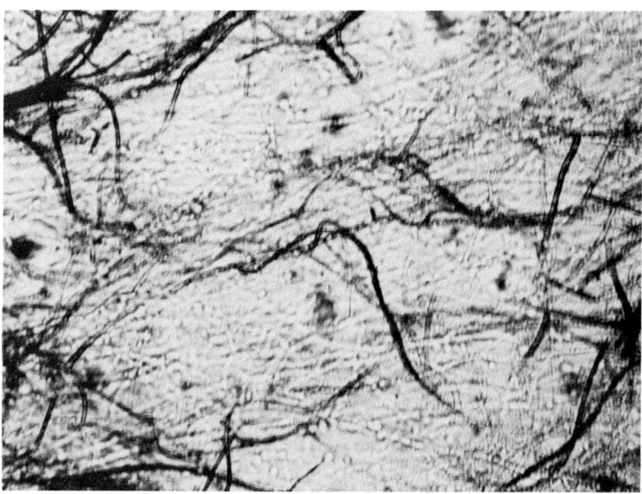

Figure 42. Blackened mycelial strands of *Verticillium albo-atrum*. Microsclerotia are not produced. (Courtesy I. Isaac)

2) Plant debris-free seed or treat with a seed protectant.

3) Mow the most recently planted alfalfa fields before mowing older plantings.

4) Disinfest machinery, footwear, etc., before leaving a site of infection.

Selected References

GONDRAN, J. 1976. Répartition en France de la Verticilliose de la luzerne. Ann. Phytopathol. 8:203-212.

GRAHAM, J. H., R. N. PEADEN, and D. W. EVANS. 1977. Verticillium wilt of alfalfa found in the United States. Plant Dis. Rep. 61:337-340.

HEALE, J. B., and I. Isaac. 1963. Wilt of lucerne caused by species of Verticillium. IV. Pathogenicity of V. albo-atrum and V. dahliae to lucerne and other crops; spread and survival of V. albo-atrum in soil and in weeds; effect upon lucerne production. Ann. Appl. Biol. 52:439-451.

ISAAC, I. 1957. Wilt of lucerne caused by species of Verticillium. Ann. Appl. Biol. 45:550-558.

SMITH, H. C. 1965. The morphology of Verticillium albo-atrum, V. dahliae, and V. tricorpus. NZ J. Agric. Res. 8:450-478.

Fusarium Wilt

Fusarium wilt of alfalfa was first described in the USA in 1927. Since then it has been observed in alfalfa-growing areas throughout the world. The disease apparently is more severe in warm areas, and although it occurs throughout the USA, it is more severe in the southern and southwestern states.

Symptoms

Wilting shoots are the first evidence of the disease. In early stages the leaves may wilt during the day and regain turgidity at night. Bleaching of the leaves and stems follows, and a reddish tinge often develops in the leaves. Only one side of a plant may be affected at first, and after several months, the entire plant dies.

The characteristics of the internal root discoloration are distinctive. Dark or reddish brown streaks occur in the stele, appearing in cross section as small partial or complete rings. In advanced stages of the disease, the outer ring of the stele or the entire stele may be discolored, resulting in death of the plant (Plate 35).

Fusarium wilt can be distinguished from bacterial wilt by the dark or reddish brown discoloration in the stele in contrast to the light or yellowish brown discoloration associated with bacterial wilt. The discolored tissue also is more sharply defined in Fusarium wilt. In Fusarium wilt the cortex or bark usually remains unaffected.

Causal Organism

Fusarium oxysporum Schlecht f. sp. *medicaginis* (Weimer) Snyd. & Hans. is the usual incitant of Fusarium wilt of alfalfa, but *F. oxysporum* f. sp. *vasinfectum* (Atk.) Snyd. & Hans. races 1 and 2 and *F. oxysporum* f. sp. *cassia* Armst. & Armst. reportedly cause similar symptoms in alfalfa.

The fungus grows rapidly on most artificial media. Cultures are felted to floccose with hyaline hyphae. Color in culture varies from colorless through pale salmon to violet or slate, depending on the medium and temperature. Optimum temperature for growth is about 25°C.

Hyaline, usually nonseptate, microconidia are always present, occasionally with bicellular conidia. Microconidia are oval to ellipsoid or cylindrical and $5-12 \times 2.2-3.5$ μm in size. They are borne on lateral phialides or short lateral conidiophores (Fig. 43).

Macroconidia are hyaline, falcate, and more or less pointed at each end. Spores usually have three septa but occasionally have four or five (Fig. 43). Macroconidia, $4-5.5 \times 25-50$ μm, are produced on lateral phialides or in sporodochia. Chlamydospores ranging from 7 to 11 μm in diameter are intercalary or terminal on mycelium. They occur singly or in short chains.

Disease Cycle and Epidemiology

The fungus survives in the soil as chlamydospores and as mycelium in live tissue and probably in plant debris. Soil may remain infested almost indefinitely.

The fungus infects small roots or enters through wounds in the taproot and progresses through the water-conducting elements of the stele. Formation of by-products or possible production of a toxic substance by the pathogen, or both, plugs the vessels, which causes wilting. Discoloration of tissue often occurs in advance of the hyphae. As the disease progresses, the plants wilt and die.

The disease usually progresses slowly in alfalfa stands, and only scattered plants show symptoms at any one time. Considerable stand loss may occur over several years, however. When the roots of susceptible plants are inoculated by dipping in a concentrated conidial suspension, the plants die within five weeks. The disease is favored by relatively high soil temperatures and is more severe in warm areas of the world.

Control

Because the pathogen persists in the soil for a long time, the only practical control is use of resistant alfalfa cultivars. A number of cultivars, especially those developed in the southwestern USA, contain a high

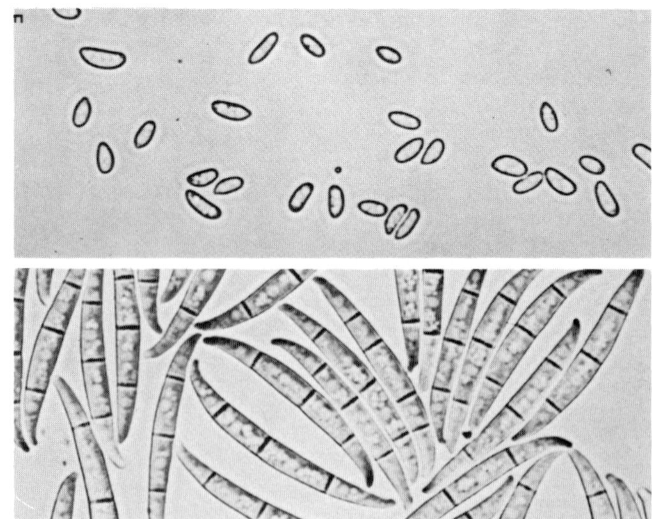

Figure 43. *Fusarium oxysporum* conidia. Microconidia (top) and macroconidia (bottom) (× 850). (Reprinted, by permission, from A Pictorial Guide to the Identification of Fusarium Species, 2nd ed., by T. A. Toussoun and P. E. Nelson. ©1976, Pennsylvania State University Press, University Park)

percentage of resistant plants. Alfalfa cultivars adapted to the northern area range from almost completely susceptible to moderately resistant.

Selected References

ARMSTRONG, G. M., and J. K. ARMSTRONG. 1965. Further studies on the pathogenicity of three forms of Fusarium oxysporum causing wilt of alfalfa. Plant Dis. Rep. 49:412-416.

FROSHEISER, F. I., and D. K. BARNES. 1978. Field reaction of artificially inoculated alfalfa populations to the Fusarium and bacterial wilt pathogens alone and in combination. Phytopathology 68:943-946.

WEIMER, J. L. 1928. A wilt disease of alfalfa caused by Fusarium oxysporum var. medicaginis. J. Agric. Res. 37:419-433.

Phymatotrichum Root Rot

Phymatotrichum root rot, also known as cotton root rot, Texas root rot, and ozonium root rot, occurs only in the region from Texas to southern California in the southwestern USA and in northern and central Mexico (Fig. 44). About 2,300 species of dicotyledonous plants are susceptible during the summer months but usually not in the winter. Monocotyledonous plants are not affected. Most researchers think that the causal agent, *Phymatotrichum omnivorum,* was an indigenous pathogen on native flora.

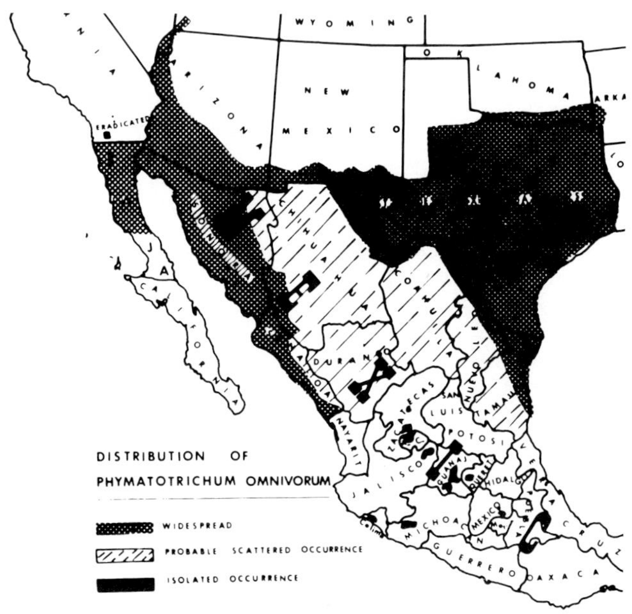

Figure 44. Distribution of *Phymatotrichum omnivorum* in the southwestern USA and in Mexico. (Reprinted, by permission, from R. B. Streets and H. E. Bloss. 1973)

Symptoms

A cortical rot of the roots leads to yellowing and bronzing of leaves and eventually to sudden and permanent wilting. If the lesions encompass the root or crown, the plant will die (Plate 36). Diseased plants may occur in a "fairy ring" pattern in which the plants at the outer edge of a circle are killed as the fungus moves out from the center of infection (Plate 37). In the center of the ring nonsusceptible grasses replace the alfalfa.

Causal Organism

Phymatotrichum omnivorum (Shear) Dug. (syn. *Ozonium omnivorum* Shear)

The fungus is classified among the Fungi Imperfecti. Mycelium is brown and consists of large cells and fine cells. The mycelium often aggregates in a strand made of many hyphae. Frequently these strands can be seen with a hand lens on the surface of an infected root. Cruciform (cross-shaped) hyphae also occur in and on the root (Fig. 45). When weather is humid, brown spore mats form on the surface of the soil on which numerous single-celled globose to ovate conidia (4.8–5.5 μm in diameter) form (Fig. 46). The function of these spores is not known because their germinability is erratic. Sclerotia (1–5 mm in diameter) are round to irregular and form on mycelium growing on affected roots (Fig. 47).

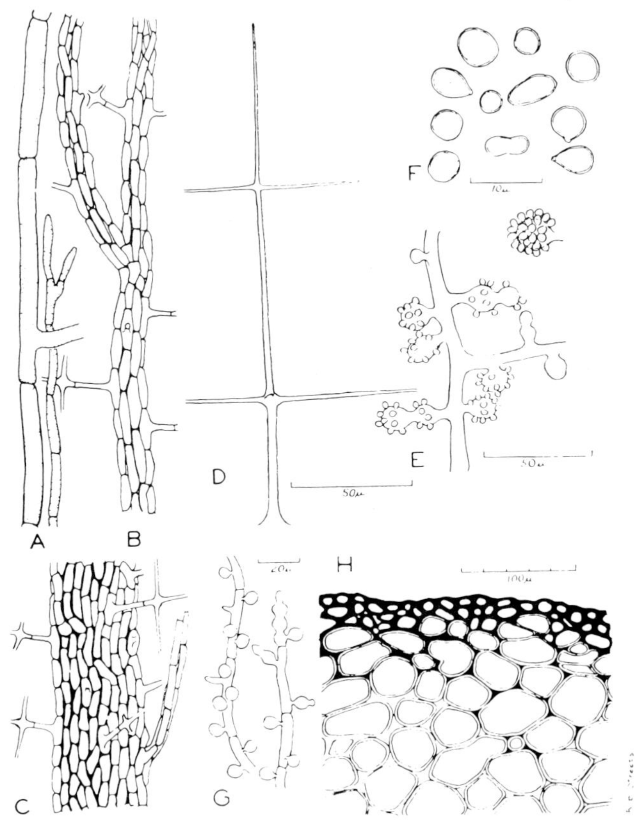

Figure 45. Morphology of *Phymatotrichum omnivorum*. **A**, Single hypha. **B**, Strand formed by branching hypha surrounding a large central hypha. **C**, Mature strand with cruciform hyphae. **D**, Cruciform hypha with acicular points. **E**, Conidia on conidiophores from a spore mat. **F**, Conidia freed from conidiophore. **G**, Conidiophores before conidiation. **H**, Cross section of a sclerotium showing thick-walled rind cells and parenchyma. (Reprinted, by permission, from R. B. Streets and H. E. Bloss. 1973)

Disease Cycle and Epidemiology

The disease occurs only in the summer months and at relatively high soil temperatures. The fungus survives as sclerotia at soil depths to 2 m. In early summer mycelium from germinating sclerotia infect plant roots. The fungus appears to be indigenous to certain fields. Although quarantines prohibit transporting soil and plants from infested areas, there is little evidence that the fungus spreads to new land. Certain calcareous soils may favor the fungus, and adding sodium chloride to these soils suppresses the disease.

Because presence of this fungus distinctly depresses the value of land, diagnosis of the disease should always be confirmed by a plant pathologist with expertise on the disease.

Control

1) Heavy application of animal manure and green manure in the form of a winter crop such as papago pea suppresses the disease. Experimental work in Texas indicates that sodium salts added to infested land also reduce disease severity.

2) Because sclerotia persist for long periods, rotation to grass crops is relatively ineffective.

3) Genetically controlled varietal resistance to the disease has not been found, but plants that vigorously produce new roots appear to have some tolerance.

Figure 46. Spore mats produced in an alfalfa field by *Phymatotrichum omnivorum*. Spores apparently have no importance, but appearance of spore mats is presumptive diagnostic evidence. (Reprinted, by permission, from R. B. Streets and H. E. Bloss. 1973)

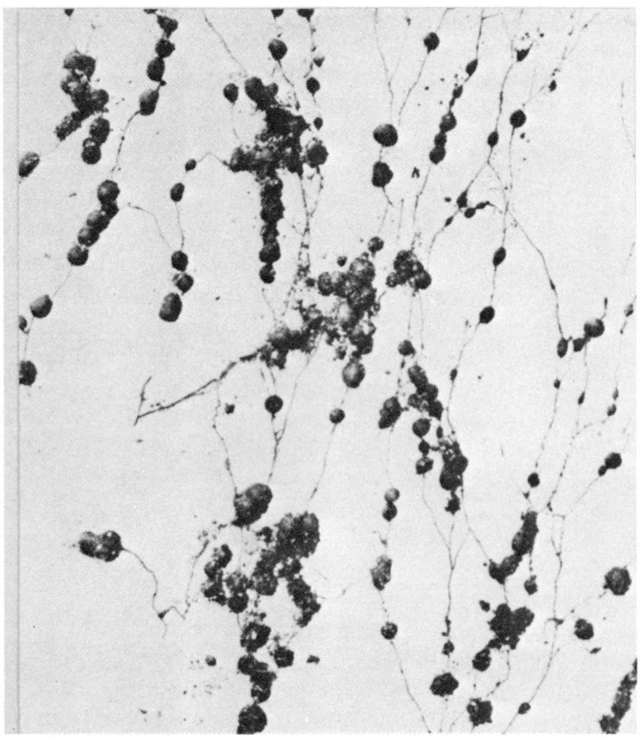

Figure 47. Sclerotia produced on roots by *Phymatotrichum omnivorum*. (Reprinted, by permission, from R. B. Streets and H. E. Bloss. 1973)

Selected References

LYDA, S. D., and E. BURNETT. 1971. Influence of temperature on Phymatotrichum sclerotial formation and disease development. Phytopathology 61:728-730.

STREETS, R. B. 1937. Phymatotrichum (cotton or Texas) root rot in Arizona. Ariz. Agric. Exp. Stn. Bull. 71:293-410.

STREETS, R. B., and H. E. BLOSS. 1973. Phymatotrichum root rot. Am. Phytopathol. Soc. Monograph No. 8. 38 p.

Winter Crown Rot or Snow Mold

Winter crown rot or snow mold is a destructive disease in western and central Alberta and northern Saskatchewan, Canada. The causal fungus, an unidentified basidiomycete, was isolated in the early 1930s. The disease may kill 50% of the plants in a stand. The severity of damage from the disease varies from year to year. The disease is less severe in southern Alberta and has been observed in Manitoba, British Columbia, and Alaska.

Symptoms

The occurrence of irregular patches of dead plants in the field is the characteristic symptom of winter crown rot in northern areas. The patches of dead plants vary in size, depending on disease severity.

Affected plants have dark brown rotted areas at the crown and less commonly on the roots. Sometimes only the crown buds and small portions of the underlying tissue are affected. Mycelium of the fungus may be seen on the soil surface in early spring. Although the crown may be completely dead, the main root may be sound until natural decay begins.

Causal Organism

An unidentified low-temperature basidiomycete that has not been observed to produce reproductive or survival structures is the causal agent. It grows in culture at 1–24°C; the optimal temperature is about 12°C. The fungus can be identified microscopically by the numerous, fully developed clamp connections on septate mycelium in colonies about two weeks old (Fig. 48). Once established, the whitish mycelium grows rather rapidly and forms a zoned, matted colony. The fungus can be cultured from superficial mycelium on freshly infected tissue at near freezing temperatures. If superficial mycelium is absent, the fungus can be isolated from the base of infected stems. Isolates require 10–14 days to produce observable growth on potato dextrose agar at 1°C. The fungus can produce hydrogen cyanide (HCN).

Disease Cycle and Epidemiology

The fungus survives as a soil inhabitant, and when conditions are favorable in the fall, it becomes active and grows in close association with the crowns of the

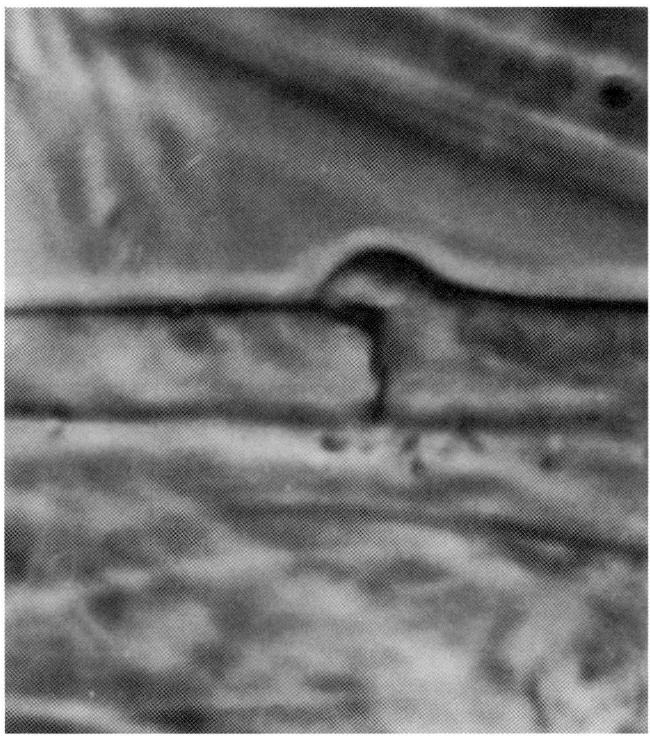

Figure 48. Clamp connection on a basidiomycetous fungus hypha. (Courtesy P. L. Lentz, USDA)

alfalfa plants.

The fungus synthesizes HCN that alfalfa crown tissue absorbs at temperatures near freezing. HCN can be detected in the crown buds about mid-December and may reach 100 ppm in February. Beginning about March 1, HCN concentration rises rapidly to 500 ppm in the crown buds. Little mycelial invasion of the host tissue occurs during the winter months until March, when extensive invasion and destruction of the crown buds occur. When temperatures rise much above freezing, the host tissue does not absorb or retain HCN and alfalfa is no longer susceptible to the pathogen.

Control

1) *Medicago falcata* is relatively resistant to winter crown rot. Cultivars with *M. falcata* parentage are more resistant than those without.

2) Rotating alfalfa with summer fallow or annual cereals or both for three seasons controls the pathogen for as long as three years in the succeeding alfalfa crop.

3) Spraying alfalfa plants twice with borax (sodium tetraborate) at 10 day intervals in the fall has prevented HCN production and given good control, but the treatment is sometimes phytotoxic.

Selected References

CORMACK, M. W. 1948. Winter crown rot or snow mold of alfalfa, clovers, and grasses in Alberta. I. Occurrence, parasitism, and spread of the pathogen. Can. J. Res. 26:71-85.

LEBEAU, J. B., and T. G. ATKINSON. 1967. Borax inhibition of cyanogenesis in snow mold of alfalfa. Phytopathology 57:863-865.

LEBEAU, J. B., M. W. CORMACK, and J. E. MOFFATT. 1959. Measuring pathogenesis by the amount of toxic substance produced in alfalfa by a snow mold fungus. Phytopathology 49:303-305.

Mycoleptodiscus Crown and Root Rot

Mycoleptodiscus crown and root rot was first observed and described in red clover (*Trifolium pratense*) in Illinois, USA, in the early 1950s. It has also been isolated from alfalfa in Illinois, Maryland, and Virginia and from birdsfoot trefoil (*Lotus corniculatus*) and soybeans (*Glycine max*) in Illinois. Other legumes including Lespedeza (*Lespedeza stipulacea*), sweet clover (*Melilotus alba*), peas (*Pisum sativum*), alsike clover (*T. hybridum*), crimson clover (*T. incarnatum*), and white clover (*T. repens*) are susceptible in artificial inoculations. The disease has been reported only in the USA.

The importance of *Mycoleptodiscus* on alfalfa is unknown, primarily because intensive surveys have not been done. The disease can often be distinguished from other root and crown rots in the field, but isolation of the causal agent may be necessary for positive identification. Alfalfa does not appear to be as susceptible as red clover and birdsfoot trefoil.

Symptoms

Mycoleptodiscus terrestris causes preemergence and postemergence damping-off of alfalfa seedlings during warm humid weather. It can be identified by the abundant sporulation on the host. On mature plants the causal fungus incites a black rot of the lateral roots and taproot and decay of the crown and interior of the taproot (Fig. 49). The black advancing margin of decay and the black sclerotia in decayed tissues help in diagnosis. Sclerotia are produced in large numbers in taproots killed by the pathogen (Fig. 50).

When foliage and stems are inoculated with a spore suspension, small leaf spots are produced and reddish brown lesions develop on the stems. The importance of leaf and stem infections under field conditions is unknown. Much of the crown rot probably results from infected stems.

Figure 49. Field-collected plant with advanced crown and root rot caused by *Mycoleptodiscus terrestris*. (Courtesy J. W. Gerdemann)

Causal Organism

Mycoleptodiscus terrestris (Gerd.) Ostazeski (first described as *Leptodiscus terrestris*)

The fungus can be isolated by planting sclerotia or tissue from the advancing margin of decay on acidified potato dextrose agar (PDA). The optimum temperature for growth in culture is near 30°C. Growth on PDA is appressed, hyaline at first and later black and crusty after sclerotia form. Conidia are produced on corn meal agar and sterilized plant tissue exposed to light.

Sclerotia are black and spherical to fusiform; they may exceed 1 mm in length in host tissue. The fruiting body is superficial on sterilized plant tissue and is yellow to dark brown; it develops radially from a central cell to form a radiating disc one cell-layer thick and 200–800 μm in diameter. Conidiophores are absent unless the disc itself is considered a highly developed conidiophore.

Conidia develop on the upper surface of the stromatic cells and form in linear rows. Successive crops of spores are produced in mucilage and adhere tightly together to form a light yellow to brown raised "spore head" in which the spores are arranged in distinct layers. The conidial walls are hyaline, but the cell contents are light yellow, often turning brown as the spores age. Conidia are usually uniseptate when mature and allantoid, with a relatively long filamentous appendage at each end. The conidia measure 20–35 × 5–7 μm; the setae are 8–18 μm long (Fig. 51).

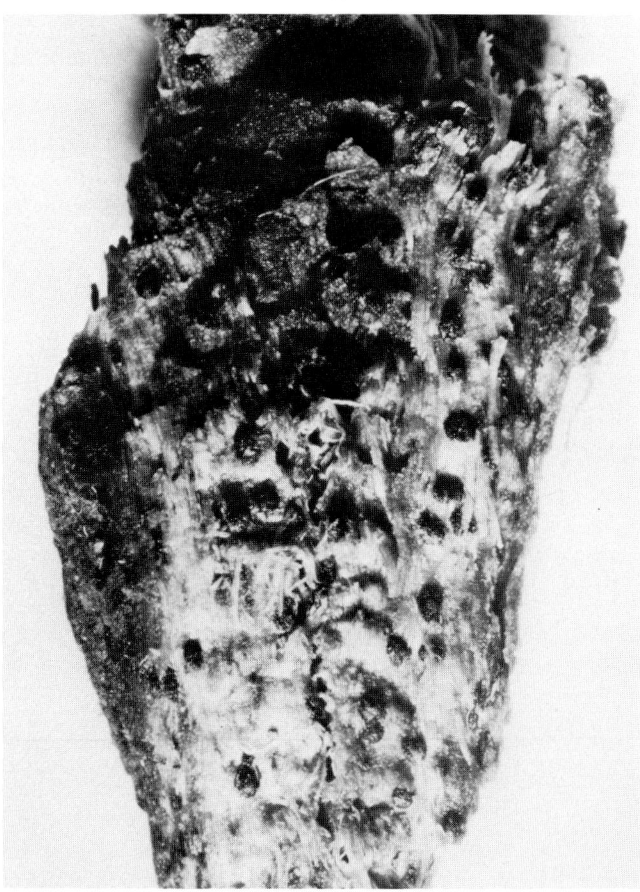

Figure 50. Close-up of taproot rotted by *Mycoleptodiscus terrestris*. Note sclerotia in decayed tissues. (Courtesy J. W. Gerdemann)

Disease Cycle and Epidemiology

The fungus probably overwinters as sclerotia in plant debris and soil. When conidia are in the mucilaginous head, the setae are folded back against the spore wall on the concave side. As the mucilage dissolves, the setae unfold and expel the spores with considerable force. Based on the optimum temperature for mycelial growth, the disease would be most damaging during the summer months.

Control

1) Control measures have not been reported.

2) The USDA has a selection program under way to develop resistant plant material.

Selected References

GERDEMANN, J. W. 1953. An undescribed fungus causing a root rot of red clover and other Leguminosae. Mycologia 45:548-554.

GERDEMANN, J. W. 1954. Pathogenicity of Leptodiscus terrestris on red clover and other Leguminosae. Phytopathology 44:451-455.

OSTAZESKI, S. A. 1967. An undescribed fungus associated with a root and crown rot of birdsfoot trefoil (Lotus corniculatus). Mycologia 59:970-975.

Cylindrocarpon Root Rot

Cylindrocarpon root rot of alfalfa was first described in 1937 after it was observed for some time in Alberta, Canada. It was recognized as one of the fungi that caused considerable stand loss in alfalfa very early in the spring. The disease has also been reported in Quebec, Ontario, and Manitoba, Canada, and recently it caused appreciable plant loss in Minnesota, USA.

Symptoms

The infected area of the root, usually just below the crown, at first has a water-soaked appearance but soon turns light brown and finally dark brown. The entire root system may rot from severe infection, but often the infected area extends only a short distance below the crown, and the remainder of the root appears sound. Sclerotialike stromata develop in the cracks of the bark of the root, giving it a rough dark appearance (Plate 38).

Causal Organism

Cylindrocarpon ehrenbergii Wr. belongs in the family Tuberculariaceae of the Fungi Imperfecti. The perfect stage, *Neonectria caespitosa* (Fckl.) Wr. has not been observed on alfalfa.

Mycelium in culture is white or a shade of orange, salmon, or violet. Sclerotialike stromata are often produced on agar. They vary in color from yellow, orange, salmon, cinnamon to brown. Conidia are produced sparingly in young cultures. In older cultures they are produced successively at the tip of the conidiophore, often adhering to each other to form a false head. The conidia are hyaline and cylindrical or slightly curved with rounded or conical ends. They are typically uniseptate and generally 20–27 × 3.3–3.9 μm

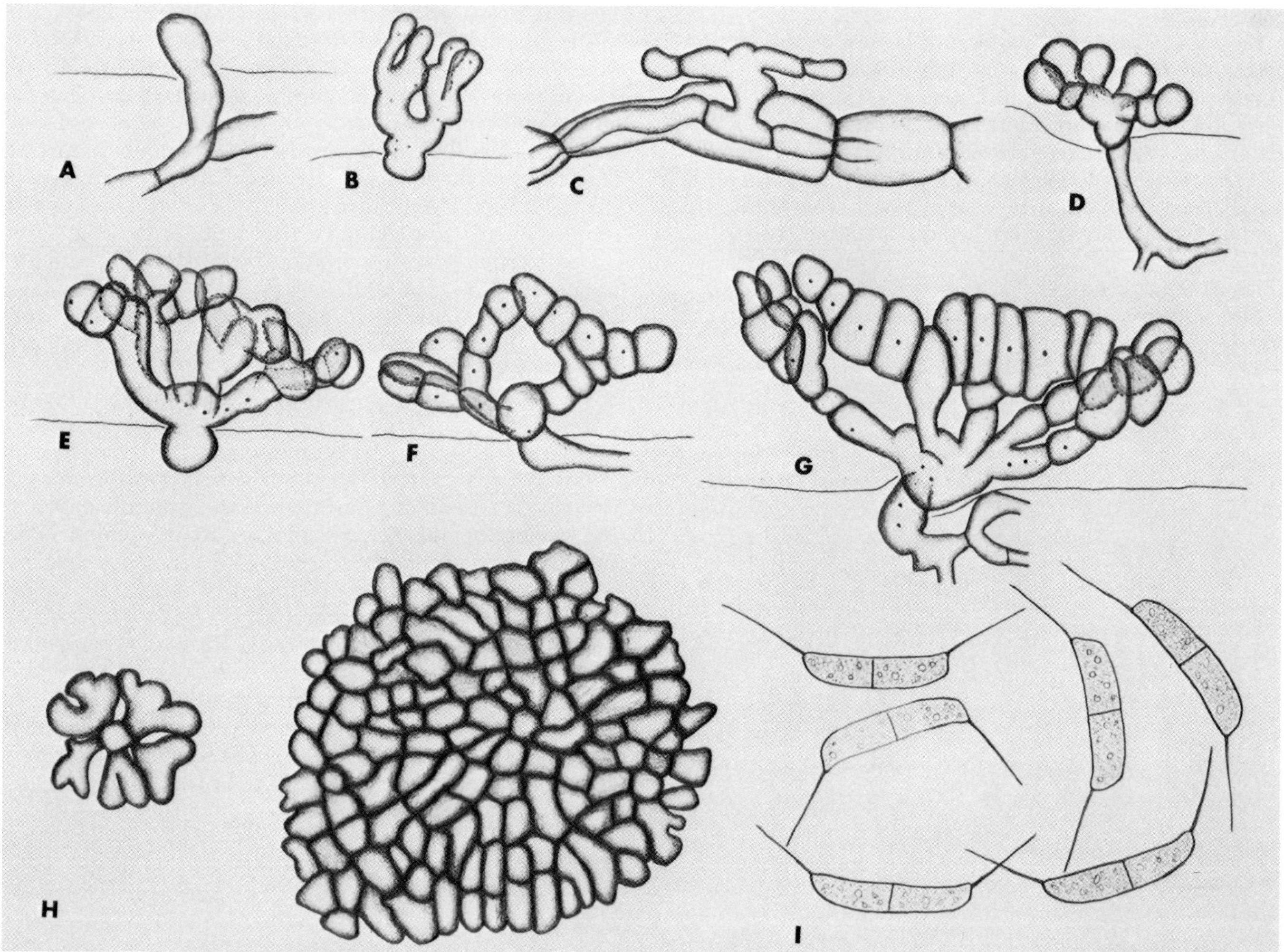

Figure 51. *Mycoleptodiscus terrestris*. **A–G**, Side views of stages in development of the fruiting structure. **A**, Initial cell formed from a hypha in an epidermal cell of the leaflet. **B**, Vertical branches arising from the initial cell. **C**, Horizontal branches arising from the initial cell. **D**, Enlarged cells at the apex of branches. **E–F**, Early stages in formation of sporogenous cells. **G**, Mature fruiting structure. **H**, Spore-bearing upper surface of a young (left) and older saucer-shaped fruiting structure, illustrating radial development from a central cell. **I**, Six conidia. (**A–G** after D. V. McVey and J. W. Gerdemann. **H–I** after J. W. Gerdemann)

(Fig. 52). Chlamydospores are usually sparse. They are spherical to ovate, intercalary, and single or in short chains and average about 10×8 μm.

C. ehrenbergii grows at -2 to 32°C. The optimum temperature is about 20°C.

Disease Cycle and Epidemiology

The fungus survives in plant debris in the soil, and in early spring when the ground begins to thaw, root infection takes place. Wounds of the roots facilitate infection but are not necessary. The pathogen can enter the roots through the tissue at the base of branch roots, through lenticels, or by direct penetration. Infection through the base of the branch roots is probably the most common. After entering the root, the hyphae develop rapidly in the phloem parenchyma.

Alfalfa plants appear to be most susceptible at the end of winter dormancy when the ground is thawing. The critical infection period may last only a few days. Later the roots become relatively resistant to invasion by *C. ehrenbergii*. Infected plants may die before regrowth is apparent in the spring.

Control

1) Resistant plants can be identified in most alfalfa populations, but no alfalfa cultivars resistant to the pathogen have been developed.

2) The fungus apparently does not attack cereal crops, and rotating cereals with alfalfa to prevent excessive buildup of the pathogen in the soil might be advisable.

Selected Reference

CORMACK, M. W. 1937. Cylindrocarpon ehrenbergi Wr. and other species, as root parasites of alfalfa and sweet clover in Alberta. Can. J. Res. 15:403-424.

Plenodomus or Brown Root Rot

Plenodomus root rot was described in 1933 but was observed as early as 1926 in Alberta and Saskatchewan, Canada. Although more severe on sweet clover, it is also destructive on alfalfa, causing considerable plant loss in early spring. It has been observed on alfalfa in Manitoba, on alsike clover in Alaska, on red and sweet clover in the Yukon, and on red clover in Finland.

Symptoms

Brown, slightly sunken necrotic lesions on the taproot or lateral roots characterize the disease. The lesions develop rapidly in all root tissues. Brown to black pycnidia usually are abundant on the surface of the lesions or may be embedded in necrotic tissue. When the rot progresses to the crown, the plant dies. If the rot is arrested while enough taproot remains, the plant will recover by producing branch roots near the crown.

Causal Organism

Plenodomus meliloti Dearn. & Sanford belongs in the Sphaeropsidaceae family in the Fungi Imperfecti.

The mycelium in mass is light gray or ashen on carbohydrate media. Young hyphae are hyaline but become brown with thicker walls as they age. The width of the mycelium varies from 2 to 5 μm and the crosswalls are 10–15 μm apart. Dark pycnidia are produced abundantly in light or darkness. They are mostly superficial but may be submerged in artificial media or in host tissue. They range from 0.5 to 2 mm in diameter. Initiation of the pycnidia follows closely the growth of the mycelium. Pycnidia are usually single on host tissue and frequently aggregated on artificial media (Fig. 53).

Ordinarily spores are produced when the pycnidia are about 50 days old and escape in a yellow exudate through the ostioles. The spores are hyaline and nonseptate and average 2.8×5.2 μm in size. The fungus grows at 0–27°C temperatures, with 15–17°C being optimum. A pH of about 6.2 appears best for the fungus.

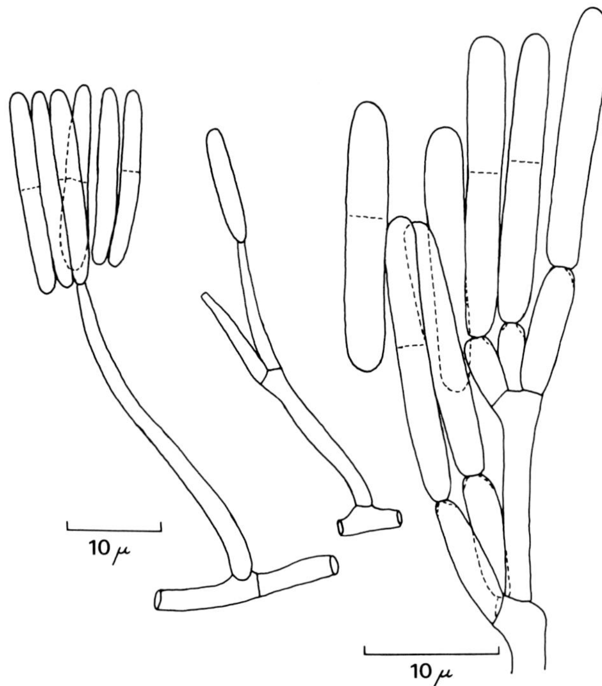

Figure 52. Conidia and conidiophores of *Cylindrocarpon ehrenbergii*. (Drawing by M. C. Shurtleff)

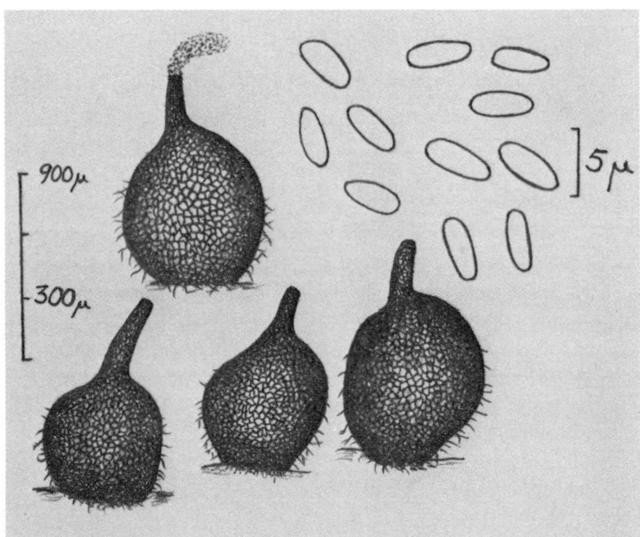

Figure 53. Pycnidia and pycnidiospores (conidia) of *Plenodomus meliloti*. (Drawing by M. C. Shurtleff)

Disease Cycle and Epidemiology

The fungus survives in plant debris in the soil and invades the roots of alfalfa when the ground thaws in the spring or during mild periods in the winter. The fungus invades nonwounded roots by direct penetration as readily as it enters through wounds; it grows intercellularly and intracellularly. Death of tissue in advance of the hyphae is common. The root system may be partially or completely rotted. As spring growth of the alfalfa progresses, the plants appear to become more resistant to the fungus, and decay is arrested. Apparently alfalfa plants are susceptible only for a short period after dormancy. The severity of the disease varies from season to season.

Control

1) No resistant alfalfa cultivars have been developed.

2) Rotating alfalfa crops with nonleguminous crops should minimize the disease incidence.

Selected References

CORMACK, M. W. 1934. On the invasion of roots of Medicago and Melilotus by Sclerotinia sp. and Plenodomus meliloti D. and S. Can. J. Res. 11:474-480.

SANFORD, G. B. 1933. A root rot of sweet clover and related crops caused by Plenodomus meliloti Dearness and Sanford. Can. J. Res. 8:337-348.

Other Crown and Root Rots

Crown and root rots are chronic disease problems of alfalfa throughout the world. Those that can be attributed to individual pathogens have been described. Frequently, however, crown or root rots cannot be ascribed to the activity of a single pathogen, and the term "complex" is used to describe the pathogens and other agents involved. The complex includes several genera, species, and strains of pathogenic fungi that interact with nonpathogenic soilborne organisms and various abiotic factors to cause rot. The nature of this complex is not yet well understood.

Symptoms

Although the symptoms caused by the root and crown rot complex are often distinctive, they are not specific to any one of the various fungi that make up the complex. Brown necrotic areas frequently are restricted to the crown or root cortex. In some instances the central core is rotted and hollow. In later stages of development, the rot may affect both cortex and vascular tissues (Plate 39). Plant vigor declines as more of the root system rots. Plants with advanced stages of rot are stunted and ultimately wilt and die. Crown rot often causes plants to develop asymmetrically, due to the death of crown bud meristem in a portion of the crown. New lateral roots that originate near the crown sometimes sustain a plant temporarily after the taproot dies.

Causal Organisms

Determining the specific organisms causing root and crown rot is extremely difficult because the same organisms may be associated with both healthy and diseased plants. In addition, fungi isolated from diseased areas may exhibit pathogenic capabilities only under specific conditions or only in specific associations.

Fusarium spp. are isolated most frequently, but many that are isolated from root or crown tissue are not pathogenic. Among the pathogenic species, *F. oxysporum* Schlecht., *F. solani* (Mart.) Appel & Wr., and *F. roseum* Lk. ex Fr. emend. Snyd. & Hans. are isolated most frequently. In addition to those previously discussed, however, fungi such as *Pythium, Rhizoctonia,* and *Phoma* spp. have also been isolated from rotted root and crown tissue.

The use of different isolation methods in ecologically different parts of the world contributes to the great variation reported in isolates from plants affected by crown rot.

Disease Cycle and Epidemiology

Fusarium spp. persist as chlamydospores in the soil or in dead plant material. Roots, which exude nitrogenous substances, stimulate the germination of chlamydospores and growth of these fungi. Roots may be colonized very early in the life of the plant. The fungi penetrate the roots either directly or through wounds and colonize the inner root tissue, often without causing rot symptoms. Small roots may rot quite rapidly, but usually rot develops slowly in taproots and crowns, often over months or years.

Various plant stresses increase the rapidity of the root rot development. Foliar diseases, foliar and root insects, frequent or untimely harvests, early frosts, poor fertility, severe winter conditions, low light intensity, and low soil pH are examples of stresses known to enhance root rot development. Stresses often interact to cause even greater injury, e.g., plants that have root rot and are stunted by leafhopper feeding may not die until they are further stressed by winter conditions. The full effect of root rot that develops during one growing season may not be evident until the following spring when plants fail to recover.

Rhizoctonia solani, a universal soil inhabitant, alone or combined with *Fusarium, Phoma,* or other species, invades the crown bud tissue, which becomes necrotic. As the plants become older, more crown buds are attacked and most or all of the meristematic tissue of the crown may be killed after several seasons (Fig. 54).

The crown rot, which is often characterized as dark brown to black wedge-shaped necrotic tissue in the center of the crown or as longitudinal lesions on the crown branches, eventually kills the plant. It is probably caused by a complex of pathogens associated with adverse physiologic or mechanical factors (Plate 40).

Figure 54. Crown rot. May be caused by several fungi. Most of the meristematic tissue has been killed. (Courtesy C. Richard)

The recurrent traffic over alfalfa fields during harvest mechanically damages the crown tissue and undoubtedly contributes to increased crown rot and alfalfa decline.

Control

Use sound management practices to minimize losses caused by the crown and root rot complex:

1) Maintain proper cutting schedule.

2) Maintain adequate soil fertility levels, especially of potassium.

3) Avoid mechanical injury as much as possible.

4) Control insects.

Selected References

ERWIN, D. C. 1954. Relation of Stagonospora, Rhizoctonia, and associated fungi to crown rot of alfalfa. Phytopathology 44:137-144.

HAWN, E. J. 1958. Studies on the epidemiology of crown bud rot of alfalfa in southern Alberta. Can. J. Bot. 36:239-250.

LEATH, K. T., and R. A. BYERS. 1977. Interaction of Fusarium root rot with pea aphid and potato leafhopper feeding on forage legumes. Phytopathology 67:226-229.

LEATH, K. T., F. L. LUKEZIC, H. W. CRITTENDEN, E. S. ELLIOTT, P. M. HALISKY, F. L. HOWARD, and S. A. OSTAZESKI. 1971. The Fusarium root rot complex of selected forage legumes in the Northeast. Pa. Agric. Exp. Stn. Bull. 777, 64 p.

O'ROURKE, C. J., and R. L. MILLAR. 1966. Root rot and root microflora of alfalfa as affected by potassium nutrition, frequency of cutting, and leaf infection. Phytopathology 56:1040-1046.

(Prepared in most part by K. T. Leath)

Additional Fungi Associated With Alfalfa

A number of fungi, in addition to those described previously, have been reported to be associated with alfalfa. Because information on their pathogenicity, distribution, and importance is lacking, they are merely listed below.

TABLE I. Additional Fungi Affecting Alfalfa

Fungus	Symptoms	Location
Acrostalagmus sp.	...	...
Aphanomyces spp.	Root rot	Ohio, Wisconsin, USA; Canada
Armillaria melea (Vahl ex Fr.) Kummer	Mushroom rot	California, USA
Botrytis cinerea Pers. ex Fr. (Fig. 55)	Shoot and flower blight (Fig. 56)	Pennsylvania, Wisconsin, USA; Japan; South Africa; USSR; Poland
Botryodiplodia sp.	Root rot	Mississippi, USA; Italy
Calonectria theae Loos (*Cylindrocladium* sp.)	Root and crown rot	India
Diplodia sp.	Crown rot	Arizona, USA; France
Erysiphe polygoni DC. ex St. Amans	Powdery mildew	Massachusetts, Wyoming, USA; Italy
Leveillula taurica (Lév.) Arn.	Powdery mildew	Japan; Sudan; USSR
Oidiopsis sp.	Powdery mildew	Brazil; Asia
Fuligo septica (L.) Wiggers	Slime mold	Maryland, USA
Bipolaris sorokiniana (Sacc. in Sorok.) Shoemaker	Stem blackening	Minnesota, USA
Microdiplodia medicaginis Died.	...	...
Macrophomina phaseolina (Tassi) G. Goid.	Charcoal rot	Illinois, Oklahoma, USA; Australia; Pakistan; Uganda, East Africa; USSR
Neocosmospora vasinfecta E. F. Sm.	...	California, USA; South Africa
Ophiobolus tenellus (Auersw.) Sacc.	On stems	Maryland, Virginia, USA
Physarum cinereum (Batsch) Pers.	Gray mold (slime mold)	Worldwide
Rhizoctonia leguminicola Gough & Elliott	Blackpatch	West Virginia, USA
Pyrenochaeta sp.	...	Canada; Hungary
Rosellinia necatrix Prill.	White root rot	California, USA; Kenya
Thielaviopsis basicola (Berk. & Br.) Ferr. (Fig. 57)	Root rot	Maine, Maryland, New Jersey, Texas, USA
Volutella spp.	On foliage	Iowa, USA; Czechoslovakia

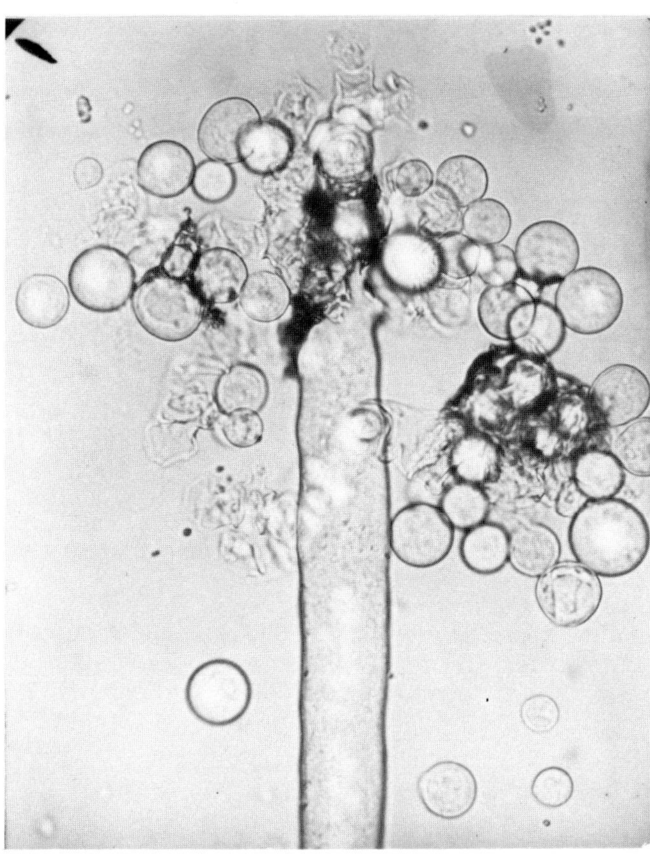

Figure 55. Conidiophore and conidia of *Botrytis cinerea*. (Courtesy N. Nishihara)

Figure 56. Shoot blight caused by *Botrytis cinerea*. (Courtesy N. Nishihara)

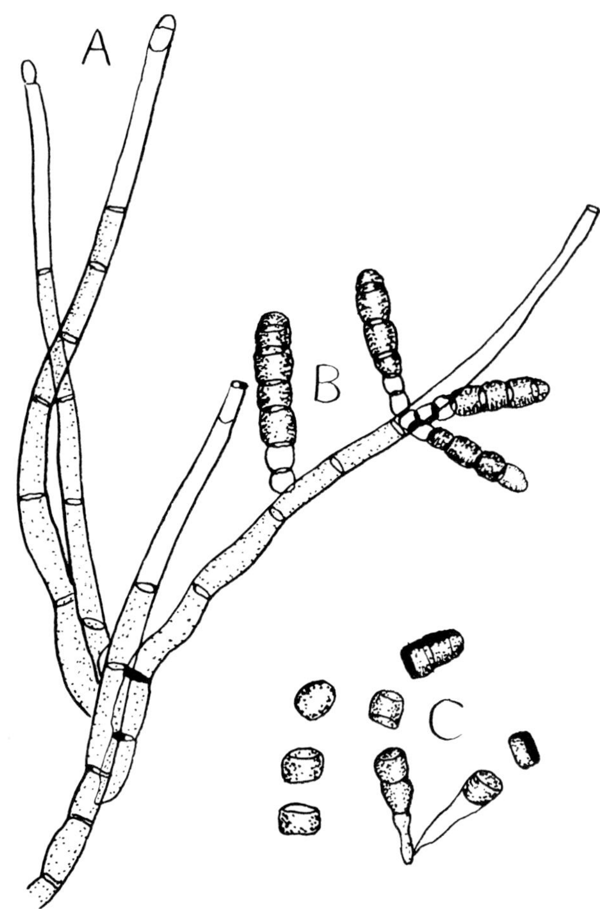

Figure 57. *Thielaviopsis basicola*. **A,** Endoconidial cells and emerging endoconidia. **B,** Chlamydospores. **C,** Chlamydospores breaking into segments. (After W. W. Gilbert)

Virus and Mycoplasmalike Agents

A number of virus diseases of alfalfa have been described, but relatively little is known about their distribution and importance in alfalfa, with the exceptions of alfalfa mosaic virus, alfalfa enation virus, and lucerne transient streak virus.

Among the viruses reported on alfalfa are alfalfa mosaic virus, pea enation virus, alfalfa enation virus, alsike clover mosaic virus, clover yellow mosaic virus, bean yellow mosaic virus, pea leafroll virus, pea streak virus, tobacco ringspot virus, white clover mosaic virus, alfalfa latent virus, and lucerne (alfalfa) transient streak virus.

Selected Reference

1970–19- -. Descriptions of plant viruses. Series issued jointly by the Commonw. Mycol. Inst. and Assoc. Appl. Biol.

Alfalfa Mosaic

Alfalfa mosaic was first described in 1931 and now is known to occur worldwide. Most alfalfa stands have infected plants, and 80% or more infected plants in stands older than two years is not uncommon.

Symptoms

The alfalfa mosaic virus (AMV) complex consists of many strains that differ in infectivity and symptomatology. Alfalfa populations also include many genotypes that react differently to AMV strains. Because of the diversity in both the pathogen and the host, the symptoms of AMV infection in alfalfa range from mild and almost imperceptible to severe, causing death of the plants.

The classic symptom is an interveinal light green or yellow mottle that may be accompanied by stunting (Plate 41). Other symptoms are stunting with several types of leaf and petiole contortions that may be combined with mottling or yellowing of leaves. Some strains cause root necrosis and death of certain susceptible alfalfa genotypes within several weeks after inoculation. Other strains have a debilitating effect, causing infected plants to die after several seasons. Severe symptoms usually are not obvious in alfalfa stands because the canopy hides severely infected plants and competition eliminates them. Symptoms of AMV are often masked during warm weather and are apparent only in the spring, probably because of cool temperature, high light, and lush growth. Most AMV-infected plants in an alfalfa stand never show symptoms.

Causal Agent

The AMV complex is composed of many strains that differ in their infectivity and other characteristics. Strains of this virus are reported to infect plants in at least 220 species in 73 genera, and the list continually increases. Alfalfa appears to be host to more strains than any other plant species.

The virus particles are of two general types. The most commonly observed are bacilliform, 16–22 nm wide, and 36–60 nm long (Fig. 58A). Some strains in certain tissues are characterized by flexuous rods with rounded ends of about the same diameter and more than 2,000 nm long (Fig. 58B). The physical characteristics such as thermal inactivation, dilution end point, and longevity in vitro are so varied that they are of little value in identification.

The virus is readily sap transmitted, but aphids are primarily responsible for transmission in the field. The chief vector in alfalfa is the pea aphid, *Acyrthosiphon pisum* (Harris). AMV is also transmitted through alfalfa seed. Field-grown alfalfa seed may be as much as 10% infected, and a 1–3% level of infected seeds is common in commercial seed lots. The virus is transmitted at a much greater frequency through pollen than through ovules. The frequency of seed transmission depends on the alfalfa genotype, AMV strain, and temperature. The virus is known to persist in alfalfa seeds for at least five years.

Disease Cycle and Epidemiology

Infected seeds are the most likely primary source of initial AMV infection in the field. Vectors also may introduce the virus from nearby infected plants. The virus spreads in the field chiefly by the pea aphid. The spotted alfalfa aphid *Therioaphis maculata* (Buckton)

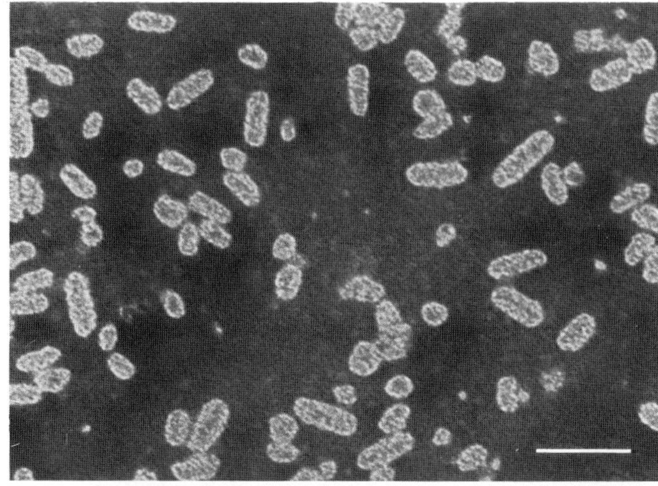

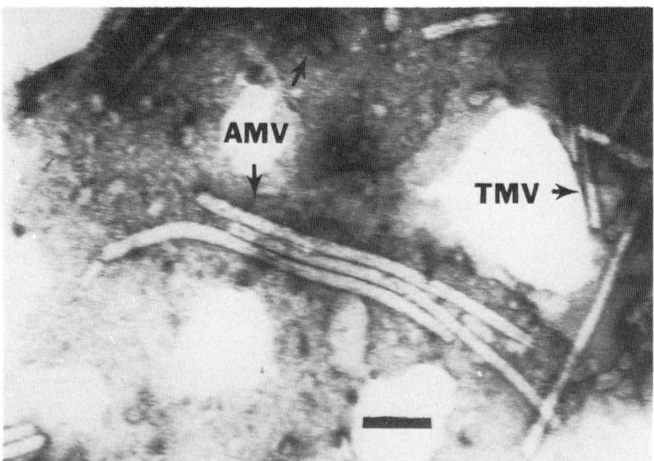

Figure 58. Alfalfa mosaic virus (AMV) particles. (Top) Bacilliform-type particles; bar represents 100 nm. (Courtesy P. R. Desjardins) (Bottom) Flexuous rod particles of AMV and tobacco mosaic virus (TMV) particles; bar represents 100 nm. (Courtesy L. E. B. Johnson)

and the blue alfalfa aphid *Acyrthosiphon kondoi* (Shinji) may be important vectors, but their ability to transmit AMV is unproved. Mechanical transmission between or in fields by harvesting machines is a possibility but has not been demonstrated.

An alfalfa plant remains infected throughout its life. The virus has been observed in the cytoplasm of cells in parenchymatous tissue in all organs except the unfertilized ovules. Limited AMV transmission through the ovule indicates that the virus is also present in some unfertilized ovules.

AMV reduces seed yield in the glasshouse, but data on reduction of seed yield due to AMV in the field are lacking. Experimental data indicate that AMV reduces forage yield in commercial fields.

Control

1) Because initial field infection can result from infected seeds, the use of virus-free seed is the first step in control of AMV. Plants used to produce first generation seed must be free of virus, and adequate insect control must be practiced in succeeding generations of alfalfa seed production.

2) Control of insects, especially aphids, in forage fields may reduce the rate of virus spread.

3) In general, all alfalfa cultivars are susceptible to AMV, but they differ in their degree of susceptibility. About 20% of the plants in most cultivars have resistance to certain AMV strains. The many prevalent AMV strains and the heterozygous nature of alfalfa complicate breeding of AMV-resistant cultivars.

Selected References

CRILL, P., D. J. HAGEDORN, and E. W. HANSON. 1970. Alfalfa mosaic, the disease and its virus incitant. Univ. Wis. Res. Bull. 280. 40 p.

FROSHEISER, F. I. 1969. Variable influence of alfalfa mosaic virus strains on growth and survival of alfalfa and on mechanical and aphid transmission. Phytopathology 59:857-862.

FROSHEISER, F. I. 1974. Alfalfa mosaic virus transmission to seed through alfalfa gametes and longevity in alfalfa seed. Phytopathology 64:102-105.

HULL, R. 1969. Alfalfa mosaic virus. Adv. Virus Res. 15:365-433.

WILCOXSON, R. D., L. E. B. JOHNSON, and F. I. FROSHEISER. 1975. Variation in aggregation forms of alfalfa mosaic virus strains in different alfalfa organs. Phytopathology 65:1249-1254.

Alfalfa Enation

A serious virus disease, alfalfa enation is widespread in southern France. It is similar, if not identical, to an enation disease of alfalfa in Spain and the virus papillosity reported from Romania. Alfalfa enation was recently identified in Bulgaria, Romania, Switzerland, and Morocco.

Symptoms

Enations or papillae, as long as several millimeters, occur on the underside of the midvein of the leaflets, which are crinkled. Plants may attain normal size but are bushy. The phloem cells are flat and laminated and the phloem parenchymal cells are hypertrophied (Plate 42).

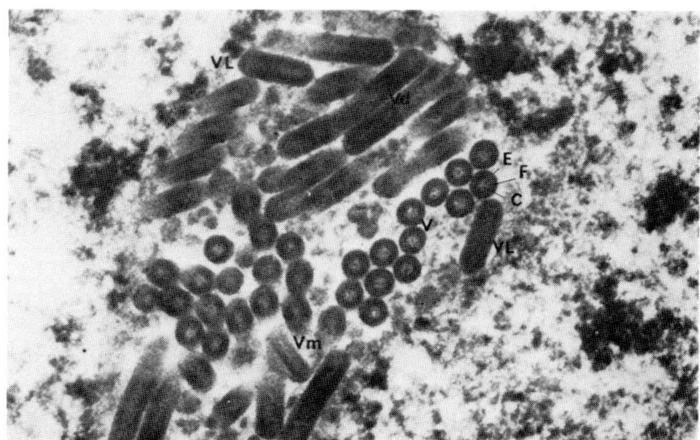

Figure 59. Alfalfa enation virus particles. **E,** External membrane, 9–10 nm thick. **F,** Internal membrane surrounding **C,** central zone, which is less electron dense. **VL,** Individual virus particles. **Vd,** Particles attached end to end. **V,** Individual virus particles in cross section. **Vm,** Particles in small numbers, grouped in vesicles surrounded by membrane. (Courtesy Bernard Alliot)

Causal Agent

The alfalfa (lucerne) enation virus is found in the nucleus and between the perinuclear membrane in companion cells and phloem parenchymal cells. The virus particles are bacilliform and are 82–89 nm in diameter and about 250 nm long. Cross sections show a central core surrounded by a thin membrane, which is in turn surrounded by an envelope or membrane 8–9 nm thick (Fig. 59).

Disease Cycle and Epidemiology

The virus is not transmitted mechanically, but it can be transmitted by grafting and is transmitted in the field by the cowpea aphid, *Aphis craccivora* Koch. The aphids transmit the virus after a 48-hr acquisition period, which identifies it as a circulative type. Symptoms appear about nine days after inoculation. The pea aphid, *Acyrthosiphon pisum* (Harris), does not transmit the virus. The virus is probably not transmitted through seed.

Control

Control methods have not been established.

Selected References

ALLIOT, B., J. GIANNOTTI, and P. A. SIGNORET. 1972. Mise en évidence de particules bacilliformes de virus associés à la maladie à énations de la Luzerne. C. R. Acad. Sci. Paris, Sér. D. 274:1974-1976.

ALLIOT, B., and P. A. SIGNORET. 1972. La maladie à énations de la Luzerne, une maladie nouvelle pour la France. Phytopathol. Z. 74:69-73.

LECLANT, F., B. ALLIOT, and P. A. SIGNORET. 1973. Transmission et épidémiologie de la maladie à énations de la Luzerne (Lev). Premiers résultats. Ann. Phytopathol. 5:441-445.

Transient Streak

Transient streak in alfalfa was recently observed and studied in Australia. It was widely distributed in the major alfalfa-growing areas in Victoria.

Symptoms

Chlorotic streaks are centered on the main lateral veins of the leaflets and vary in size from small spots to streaks 1–2 mm wide, extending to the leaf margin (Fig. 60). Symptoms are most obvious on newly expanded leaflets that often are distorted around the streaks. The symptoms fade as the leaf ages. No symptoms occur during the summer when temperatures are high.

Causal Agent

Lucerne (alfalfa) transient streak virus particles are angular, usually hexagonal, with no apparent subunit structure. The particles are 26–28 nm in diameter (Fig. 61). The virus in crude *Chenopodium quinoa* sap has a thermal inactivation point of 55–65°C at 10 min and a dilution end point of 10^{-6}–10^{-7}. It remains infective in storage at 20°C for 53 but not 60 days.

In addition to alfalfa, two species of *Medicago, Chenopodium, Lupinus,* and *Vicia,* one species of *Cicer, Pisum,* and *Zinnia* are susceptible to sap inoculation. Transmission attempts with the aphid species *Aphis craccivora* Koch, *Acyrthosiphon solani* (Kltb.), and *Myzus persicae* (Sulz.) have been unsuccessful. The virus is not known to be transmitted by seed.

Figure 60. Leaf streak symptoms of lucerne transient leaf streak virus disease on alfalfa. (Courtesy John Blackstock, Victoria, Australia)

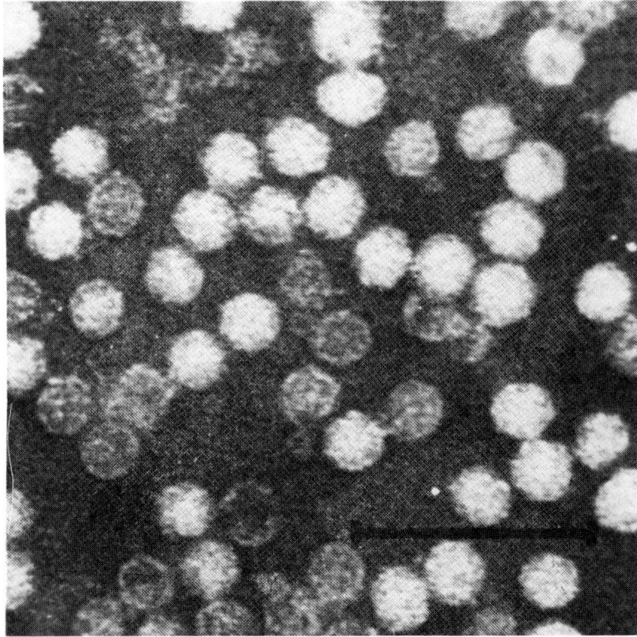

Figure 61. Particles of lucerne transient streak virus mounted in neutral phosphotungstate. Bar represents 100 nm. (Reprinted, by permission, from John Blackstock. 1978. Aust. J. Agric. Res. 29:291-304.)

Disease Cycle and Epidemiology

Lucerne transient streak virus is readily transmitted by sap but is not transmitted by seed. The method of transmission in nature is not known; insect transmission has not been proved.

Forage yield reductions of 18% in the field and 12% in the glasshouse resulted under experimental conditions when 100% of the plants were infected.

Control

No control method has been established.

Selected Reference

BLACKSTOCK, J. McK. 1978. Lucerne transient streak and lucerne latent, two new viruses of lucerne. Aust. J. Agric. Res. 29:291-304.

Alfalfa Latent Virus

Alfalfa latent virus (ALV) was recently isolated from alfalfa plants from four commercial fields in Nebraska, USA.

Symptoms

Infected alfalfa plants have no apparent symptoms. Slight reddish brown to brown necrosis develops on young leaves of infected broad bean (*Vicia faba* L.). Plants of other susceptible species are symptomless when infected with the virus under normal environmental conditions.

Causal Agent

ALV is a member of the carlavirus group. The particles are straight to slightly flexuous rods, ranging from 580 to 700 nm, with a modal length of about 650 nm (Fig. 62). ALV in crude sap has a thermal inactivation

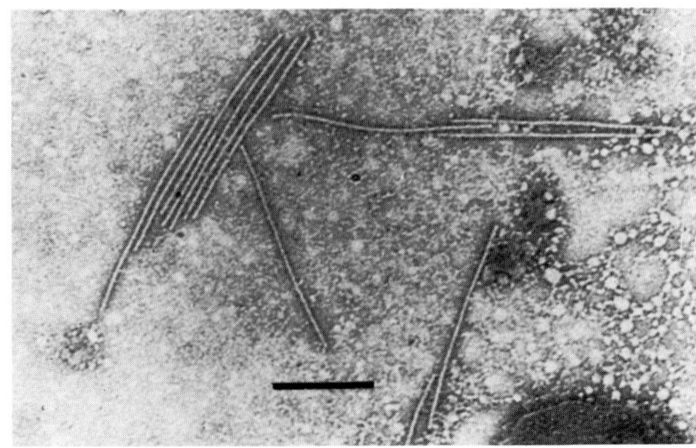

Figure 62. Electron micrograph of alfalfa latent virus in crude sap. Bar represents 275 nm. (Reprinted, by permission, from V. Veerisetty and M. K. Brakke. 1977)

point of 65–70°C at 10 min and a dilution end point of 10^{-3}–10^{-4}. The crude sap from broad bean is infectious for four days but not six days at 25°C.

The virus is readily transmitted by the pea aphid (*Acyrthosiphon pisum* Harris) and by sap inoculation. Other legume hosts susceptible to sap transmission are broad bean, pea (*Pisum sativum* L.), and hairy vetch (*Vicia villosa* Roth). There is no evidence of seed transmission.

Disease Cycle and Epidemiology

ALV is transmitted by the pea aphid, and because this vector is present in most alfalfa stands, it is the likely vector in the field. Because ALV was only recently discovered and described, no information on its distribution and effect on the alfalfa crop is available.

Control

No control method has been developed.

Selected Reference

VEERISETTY, V., and M. K. BRAKKE. 1977. Alfalfa latent virus, a naturally occurring carlavirus in alfalfa. Phytopathology 67:1202-1206.

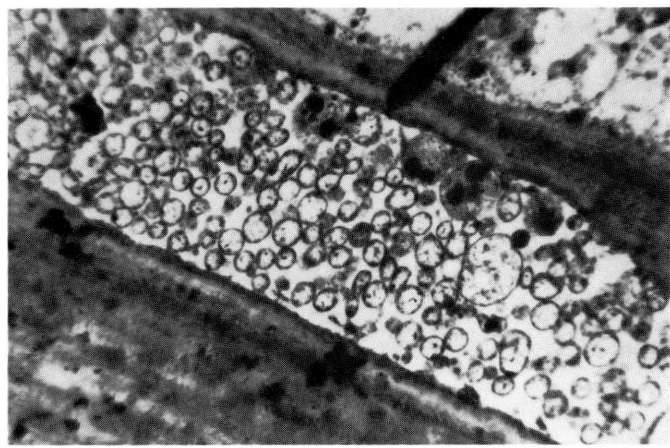

Figure 63. Mycoplasmalike bodies in sieve tubes in phloem tissue of alfalfa affected by witches' broom disease, as shown by electron microscopy of ultrathin sections. (Courtesy of John Blackstock, Victoria, Australia)

Witches' Broom

Long considered a leafhopper-transmitted virus, witches' broom is now thought to be caused by a mycoplasmalike microorganism. The disease occurs in the western USA, Canada, Australia, the USSR, and probably in other semiarid areas of the world. Except for local sporadic outbreaks, the disease is of minor importance in the USA and Canada. It is widespread in parts of Australia, however, particularly in the drier areas of New South Wales and Queensland.

The causal agent is transmitted to alfalfa and other legumes by leafhoppers, in Australia specifically by the leafhopper *Orosius argentatus* (Evans), and to nonleguminous plants by dodder (*Cuscuta* spp.). Witches' broom significantly decreases hay yield almost entirely inhibits seed yield, and gradually reduces the stand.

Symptoms

The disease is characterized by excessive development of short spindly shoots from all parts of the crown and from axillary buds along the stems. Several hundred fine stems are common on infected plants, and large, severely infected crowns may have as many as 3,000 fine, densely matted stems. Marginal chlorosis of the small, rounded, often puckered leaves and a general pale-green color of the tiny succulent stems gives the stunted plants a yellowish cast (Plate 43). Some "strains" of the causal agent produce relatively mild symptoms, however, and some infected plants have green flowers.

Infected plants may appear to recover from the disease gradually during cool periods in winter and spring, but recurrent symptoms develop rapidly with high summer temperatures accompanied by moisture stress.

Causal Organisms

Symptoms, including stunting, proliferation of axillary shoots, general chlorosis, and sometimes greening and proliferation of floral parts collectively known as "yellows," were in the past generally considered to be caused by viruses. Recent technology, however, shows that mycoplasmalike organisms—not viruses—are often the cause. Mycoplasmalike bodies are found in the phloem of alfalfa plants with witches' broom symptoms (Fig. 63), but pathogenicity according to Koch's postulates is unproved.

Control

No practical direct methods of control are known, but area-wide removal of old diseased plantings, reseeding, and maintenance of dense productive stands will reduce spread to new plantings.

Selected References

BOWYER, J. W., J. C. ATHERTON, D. S. TEAKLE, and G. A. AHERN. 1969. Mycoplasma-like bodies in plants affected by legume little leaf, tomato big bud, and lucerne witches' broom diseases. Aust. J. Biol. Sci. 22:271-274.

HELMS, K. 1957. Witches' broom disease of lucerne. II. Field studies on factors influencing symptom expression, disease incidence, and mortality rate. Aust. J. Agric. Res. 8:148-161.

Nematodal Diseases

Of the plant parasitic nematodes on alfalfa, alfalfa stem nematodes, root-knot nematodes, and root-lesion nematodes are the most important economically. Their importance is measured not only in their pathologic relationship to alfalfa but also in their importance in disease complexes involving other pathogens.

Alfalfa Stem Nematode

The alfalfa stem nematode is the most important nematode on alfalfa. This species has accompanied alfalfa seed to practically every part of the globe. It is most frequently reported as a serious pest in the Americas in heavy soils, in areas of high rainfall and heavy spring rains, and in irrigated areas. The nematode was first reported on alfalfa in Germany in 1881 and in the USA in 1923.

Causal Organism

Ditylenchus dipsaci (Kühn) Filipjev is reported in more than 300 plant hosts, and 11 distinct biologic races have been described. The alfalfa race usually is specific for alfalfa, but a Utah population of that race parasitizes several plant species but multiplies only on alfalfa or sainfoin. Males and females are 1.0–1.3 mm long. Optimum temperatures for invasion and reproduction in alfalfa are 15–20°C. A life cycle can be completed in 19–23 days in susceptible cultivars at those temperatures. Reproduction can occur from 5 to 30°C.

Symptoms, Disease Cycle, and Epidemiology

All stages except the first, which molts in the egg, can infect the alfalfa plant (Fig. 64). The nematode enters primordial bud tissue and penetrates developing buds. Infected stems enlarge and usually become discolored, the nodes swell, and the internodes are short (Plate 44). Growing stems are shortened and stunted and may succumb to the infection or may overcome the swelling and appear to grow normally. Lower stem necrosis results as the nematode continues to multiply. During long periods of parasitism accompanied by moderate temperatures and high humidity, the stem blackens for 30 cm or more above the ground. The matrix inside the blackened stem may contain several thousand nematodes per gram. The number of stems per crown are reduced as the alfalfa crown is destroyed, and eventually the entire plant dies. In some cases of heavy infestation, the nematodes migrate into the leaf, causing curling and distortion of the tissues. Some affected shoots, particularly regrowth after the first hay cutting, may be devoid of green color but only slightly smaller (Plate 45). This condition is called white flagging and is especially prominent in the Columbia River basin of Washington, USA.

Nematodes may also infect the seed if the inflorescences are infected. Roots are occasionally infected, which causes internal cavities or gall-like outgrowths.

The alfalfa stem nematode may be associated with the bacterial wilt organism, and presence of the nematode can increase the susceptibility of some alfalfa selections to bacterial wilt.

Stem nematodes congregate to form what is commonly known as "nema wool," a characteristic peculiar to this species. This phenomenon occurs in infected alfalfa tissue and in beakers of water during times of extraction.

Debris in seed was once considered the major avenue of dissemination of *D. dipsaci,* as a gram of fine screenings from uncleaned alfalfa seed may contain 37 nematodes. The nematodes are spread rapidly by machinery during the harvest season and by rain and irrigation water from alfalfa debris left in the fields. Reuse of waste irrigation water probably disseminates the nematode most effectively. In northern climates the nematode overwinters mainly in alfalfa crown tissue and in young stem tissue protected by snow cover; very few overwinter in debris and soil.

Soil moisture quickly activates this species from its quiescent stage. Survival is therefore short at high soil moisture levels in the absence of a host, but in desert areas with little or no rainfall, stem nematodes survive in fallowed soil for several years. Few or no nematodes can be recovered from fields after three years in grain crops.

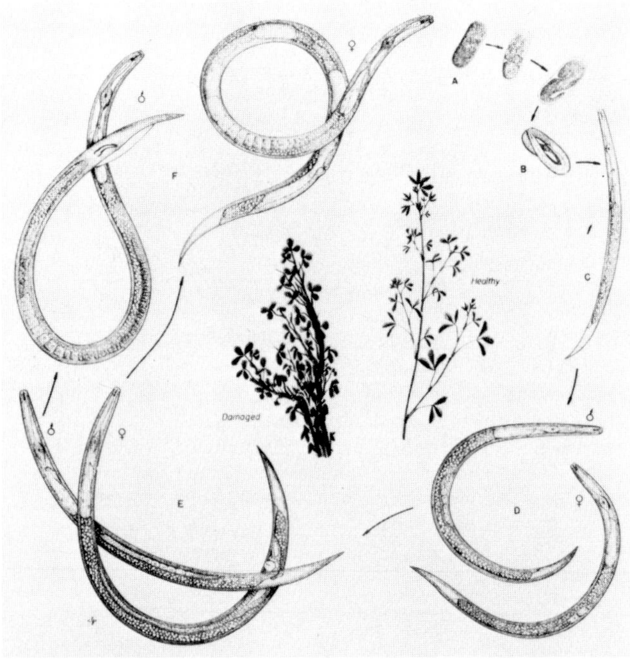

Figure 64. Life cycle of the alfalfa stem nematode *Ditylenchus dipsaci.* **A,** Egg stage. **B-C,** Second stage larvae. **D,** Third stage larvae. **E,** Fourth stage larvae. **F,** Adults. (Courtesy Department of Nematology, University of California, Riverside)

Control

1) Several alfalfa cultivars resistant to the stem nematode have been developed.

2) A two to three year crop rotation with a nonhost crop such as grain, beans, or sugar beets usually reduces the population below the detection level. However, recontamination with machinery, animals, and waste water can quickly negate the rotation effect, and the field can be reinfested in a short time.

3) In northern moderate climates, cutting the first hay crop when the top 5–8 cm of soil is very dry virtually eliminates nematode infection of the second and third hay crop. Rainfall at the time of cutting, however, may negate the effect.

4) Fall burning of alfalfa fields for weed control also decreases nematode infection the following spring. Conversely, spring burning initiates growth of new plant tissue and results in increased infection. This nematode has completely devastated two year old alfalfa stands because of spring burning.

5) Experimental trials of many systemic nematicides have been successful. Timing is of major importance. Nematicides should be applied soon after spring growth begins.

Selected References

FAULKNER, L. R., and W. J. BOLANDER. 1970. Acquisition and distribution of nematodes in irrigation waterways of the Columbia Basin in eastern Washington. J. Nematol. 2:362-367.

GRIFFIN, G. D. 1968. The pathogenicity of Ditylenchus dipsaci to alfalfa and the relationship of temperature to plant infection and susceptibility. Phytopathology 58:929-932.

HAWN, E. J., and M. R. HANNA. 1967. Influence of stem nematode infestation on bacterial wilt reaction and forage yield of alfalfa varieties. Can. J. Plant Sci. 47:203-208.

SEINHORST, J. W. 1957. Some aspects of the biology and ecology of stem eelworms. Nematologica (Suppl.)2:355-361.

THORNE, G. 1961. Principles of nematology. McGraw-Hill Book Co., Inc., New York. 553 p.

(Prepared by G. D. Griffin)

Root-Knot Nematodes

Root-knot nematodes, *Meloidogyne* spp., are probably the most widely disseminated plant parasitic nematodes in the world and rank high in economic importance as plant pathogens. Although found in most soils, root-knot species are generally favored by sandy loam soils.

Root-knot nematodes build up large populations on alfalfa but usually cause less damage to the alfalfa than to more highly susceptible crops that follow alfalfa in the rotation. They may also be involved in disease complexes with other alfalfa pathogens. *M. incognita* increases the incidence and severity of Fusarium wilt in alfalfa, and the degree of plant wilting is contingent on the severity of the nematode infection. *M. hapla* increases the incidence of bacterial wilt in both resistant and susceptible cultivars.

Symptoms

Root-knot nematode-infected plants may become stunted, and the stand may be reduced. Roots branch excessively and have small galls (Fig. 65) that are easily missed if roots are not examined carefully.

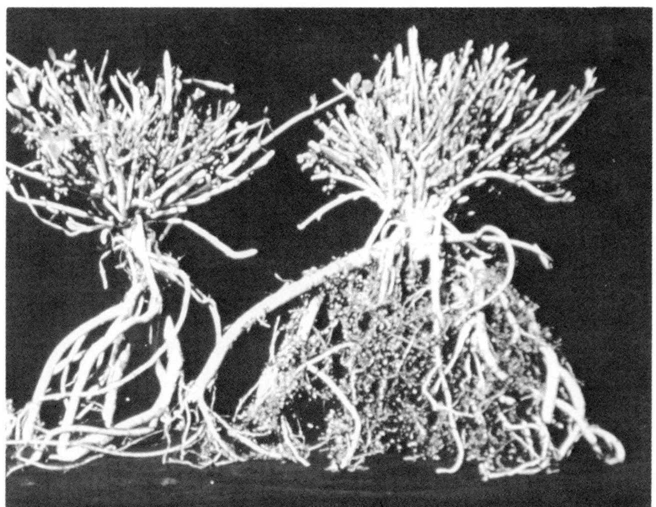

Figure 65. Healthy plant (left) and plant infected by the root-knot nematode *Meloidogyne hapla* (right), showing excessively branched roots with small galls. (Courtesy O. J. Hunt, USDA)

Causal Organism

Four common species of root-knot nematodes are important on alfalfa. They are *M. hapla* Chitwood, the northern root-knot nematode; *M. incognita* (Kofoid & White) Chitwood, the southern root-knot nematode; *M. javanica* (Treub) Chitwood, the Javanese root-knot nematode; and *M. arenaria* (Neal) Chitwood, the peanut root-knot nematode. *M. javanica* and *M. arenaria* are the least important economically.

Root-knot nematodes are extremely sensitive to soil temperatures. The northern root-knot nematode is most cold tolerant and occurs in the more northern climates, but it survives poorly in fields where the frost line extends below the feeder root zone.

Disease Cycle and Epidemiology

The infective stage of *Meloidogyne* spp. is the second stage or newly hatched larvae. The larvae are approximately 0.40–0.45 mm long. They infect only the tender plant tissue near the root tip, where their feeding causes cellular hypertrophy. The larvae become oriented toward the root plerome, feed on this tissue, and prohibit its differentiation into vascular tissue. The cells enlarge, and a cell wall may form during cell division but soon disappears. The adjoining cells coalesce, forming giant cells that become a continual food source for the nematode. Larvae become sedentary and grow. Gall size differs in relation to the root-knot species. Galls of *M. hapla* are small and oval, and those of *M. incognita* are large and more cylindrical.

Nematodes develop their second, third, and fourth body cuticles without a molt, and until the gonads form and the final molt occurs, the sexes cannot be differentiated. The female becomes oval in shape and remains sedentary throughout her life, feeding on the giant cells (Fig. 66). The male, after molting the fourth cuticle, becomes a cylindrical nematode approximately 1.2 mm long and migrates from the root. The role of the male is not entirely known since reproduction can occur without males. However, males are found in large numbers in the soil, in egg masses, and occasionally in the root tissue. There appears to be a food-sex relationship with root-knot nematodes; the ratio of males to females increases as inoculum density increases.

The female grows in size, becoming pear-shaped, and deposits a mass of 100–350 eggs in a protective gelatinous matrix that adheres to the outer surface of the root. When temperature and soil moisture are

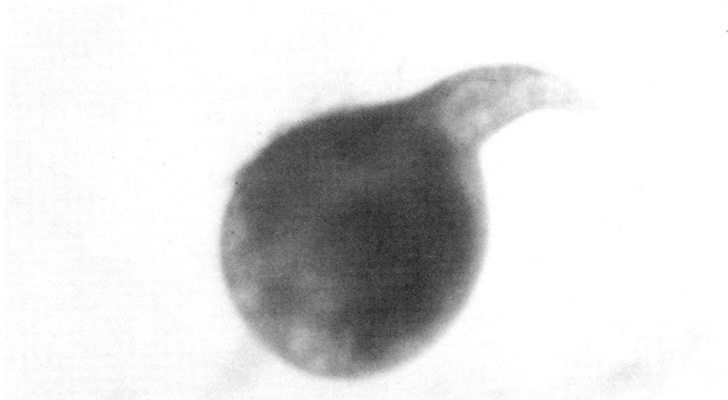

Figure 66. Mature, gravid, female root-knot nematode *Meloidogyne hapla* (× 65). (Courtesy D. H. MacDonald)

favorable, the eggs hatch and the larvae escape into the soil. The life cycle of *M. hapla* on alfalfa requires approximately 30 days at 25°C (Fig. 67).

Control

1) Resistance is the most practical method, as chemical control is not economically feasible. Some breeding lines are resistant to the northern root-knot nematode *M. hapla* and to *M. javanica*. The African type alfalfas are resistant to *M. incognita*.

2) Crop rotation does not satisfactorily control root-knot nematodes on alfalfa because they have a wide host range.

Selected References

GRIFFIN, G. D. 1969. Effects of temperature on Meloidogyne hapla in alfalfa. Phytopathology 59:599-602.

HUNT, O. J., G. D. GRIFFIN, J. J. MURRAY, M. W. PEDERSEN, and R. N. PEADEN. 1971. The effects of root-knot nematodes on bacterial wilt in alfalfa. Phytopathology 61:256-259.

SASSER, J. N. 1977. Worldwide dissemination and importance of the root-knot nematodes, Meloidogyne spp. J. Nematol. 9:26-29.

STANFORD, E. H., B. P. GOPLEN, and M. W. ALLEN. 1958. Sources of resistance in alfalfa to the northern root-knot nematode, Meloidogyne hapla. Phytopathology 48:347-349.

(Prepared by G. D. Griffin)

Root-Lesion Nematodes

Root-lesion nematodes, *Pratylenchus* spp., are cosmopolitan in distribution, occurring frequently throughout the temperate regions and the tropics. They attack a wide range of crop and weed plants and are primary factors in destroying roots of cultivated and uncultivated plants.

Several species are associated with alfalfa. *Pratylenchus* spp. have caused extensive root damage and severe reduction of spring-sown alfalfa stands in Kentucky, USA. High incidence of root-lesion nematode infection can reduce forage yields, decrease cold tolerance, and increase infections by *Fusarium* spp. Damage caused by root-lesion nematodes is difficult to evaluate accurately because other soil microorganisms usually invade and damage nematode-infected roots.

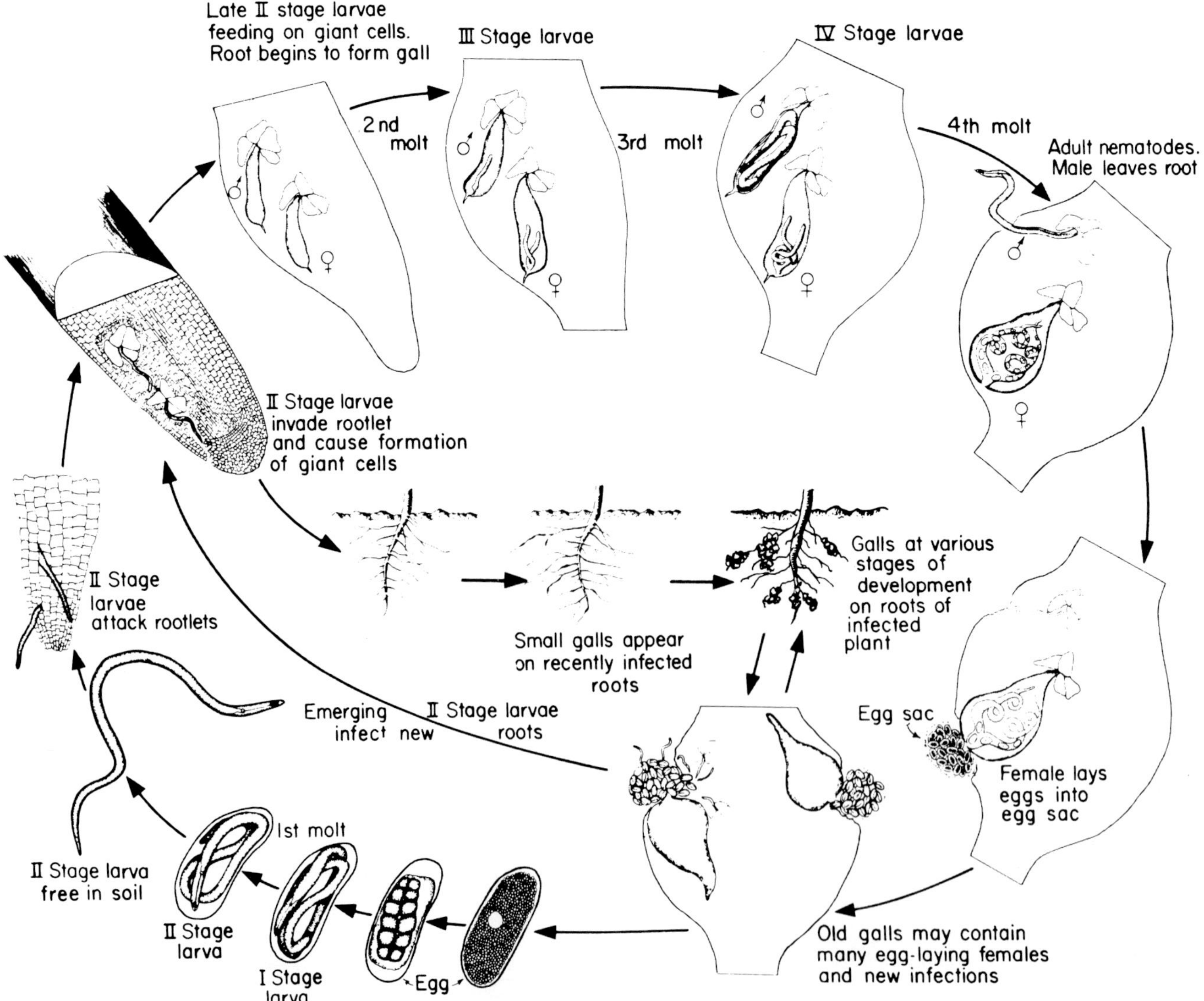

Figure 67. Disease cycle of root knot of alfalfa caused by nematodes of the *Meloidogyne* genus. (Reprinted, by permission and with slight modification, from Plant Pathology by G. N. Agrios. ©1969, The Academic Press, New York)

Symptoms

Aboveground symptoms are difficult to assess since other pathogens may produce similar symptoms. No symptoms develop when nematode numbers are low, but when numbers are high and environmental conditions are ideal, infected plants become stunted. Root-lesion nematode feeding generally causes dark lesions, overall browning of the roots, and reduced root growth.

Causal Organism

The most important *Pratylenchus* sp. associated with alfalfa is *P. penetrans* (Cobb) Filipjev & Schuurmans Stekhoven (Fig. 4). Other species include *P. crenatus* Loof, *P. neglectus* (Rensch) Filipjev & Schuurmans Stekhoven, *P. coffeae* (Zimmerman) Filipjev & Schuurmans Stekhoven, *P. pratensis* (de Man) Filipjev, *P. vulnus* Allen & Jensen, and *P. globulicola* Romaniko.

The life cycle of *Pratylenchus* is direct. Gravid females deposit eggs in infected root tissue or in the soil. Both larval and adult forms enter roots by forcing their way between or through the cortical cells and then feed on the cell contents as they migrate through the root tissue. The points of entrance to the roots provide access for soil microorganisms. Penetration of alfalfa roots by *P. penetrans* is not affected by the presence of *Meloidogyne incognita*.

Control

1) Root-lesion nematodes have extensive host ranges, and more than one species may infect a given host. Crop rotation, therefore, is of less value as a control measure for these than for other nematode-host combinations. Crops differ in their suitability as hosts for root-lesion nematodes, however, and alfalfa should not be rotated with crops that are also good hosts for the nematode.

2) Alfalfa cultivars resistant to root-lesion nematodes are not available. No useful levels of resistance have been located among *Medicago* species, but recent studies indicate that it may be possible to develop adapted cultivars with satisfactory resistance to *P. penetrans*.

3) Nematicides effectively control *P. penetrans* on alfalfa and increase forage yield, but their use is not economical and residue problems may ensue.

Selected References

CHAPMAN, R. A. 1954. Meadow nematodes associated with failure of spring-sown alfalfa. Phytopathology 44:542-545.

ROMANIKO, V. I. 1961. Injury and economic damage to legumes caused by Pratylenchus globulicola n. sp. (Trans. from Russ.). Pages 72-84 *in* Brzeski, M. W., and B. M. Zuckerman. English translations of selected Russian and Polish papers in phytonematology. 1964. Univ. of Mass. Agric. Exp. Stn. Bull. 546.

SUZUKI, M., and C. B. WILLIS. 1974. Root lesion nematodes affect cold tolerance of alfalfa. Can. J. Plant Sci. 54:585-586.

THOMPSON, L. S., and C. B. WILLIS. 1970. Effect of nematicides on root lesion nematodes and forage legume yields. Can. J. Plant Sci. 50:577-581.

TOWNSHEND, J. L., and H. BAENZIGER. 1976. Evidence of resistance to root-knot and root-lesion nematodes in alfalfa clones. Can. J. Plant Sci. 56:977-979.

(Prepared by C. B. Willis)

Other Nematodes

In addition to stem, root-knot, and root-lesion nematodes, at least 15 other plant parasitic nematodes are reported to be associated with alfalfa. These include one or more species of the genera *Aphelenchoides, Criconemoides, Hemicycliophora, Hoplolaimus, Pratylenchus, Radopholus, Tylenchorhynchus,* and *Xiphinema*.

(Prepared by G. D. Griffin)

Ineffective Nodulation

Soil sickness, probably synonymous with legume sickness or alfalfa sickness, occurs in certain fields in Canada (Alberta), the northwestern USA, and a few other areas where alfalfa is grown under continuous culture. Second year plants are often spindly or stunted and chlorotic. Many of the plants die.

Various causes have been suggested such as deficiency of essential mineral elements and unfavorable soil moisture conditions—waterlogging in early spring and drought in midsummer—associated with dense subsoil. The affected plants, however, either lack nodules or the nodules are ineffective in fixing nitrogen (Fig. 3).

The cause of the reduced symbiotic nitrogen fixation is not known. Some researchers suggest that the condition may be due to lysis of *Rhizobium* bacteria by bacteriophages or nodule induction by infective but ineffectual strains of *Rhizobium*.

Selected References

WEBER, D. F., and G. E. LEGGETT. 1966. Relation of Rhizobia to alfalfa sickness in eastern Washington. U.S. Dep. Agric. ARS 41-117. 16 p.

WEBSTER, G. R., S. U. KHAN, and A. W. MOORE. 1967. Poor growth of alfalfa (Medicago sativa) on some Alberta soils. Agron. J. 59:37-41.

Parasitic Flowering Plants

Dodder

Dodder, an annual parasitic flowering plant, is widely distributed in North America and Europe, and various species are found in the other continents. It is the most destructive weed in many alfalfa seed fields in the western United States, and where uncontrolled, dodder has ended profitable production of alfalfa seed. Dodder seed can remain dormant in the soil for as long as 20 years.

Symptoms

The first signs of the parasite are fine, inconspicuous, orange to yellow, leafless seedlings that emerge from seed in or on the soil. These attach to the host plant and produce abundant orange to yellow fine strands that grow and entwine the aboveground parts of plants. Dodder has inconspicuous flowers and tiny scales instead of leaves. The mass of entwined stems from a single plant may spread to a diameter of 9 m or more. Heavily infected alfalfa plants are weakened, matted down, and decline in vigor. Infested areas gradually assume the yellowish color of the parasitic vine (Plate 46).

Causal Organism

Three *Cuscuta* spp. commonly occur on alfalfa: *C. indecora* (large-seed dodder) is common in the western USA but rare in the east; *C. planiflora* (small-seed dodder) is severe in the west but absent in the east; and *C. campestris* (field dodder) is most damaging to alfalfa. It has a wide host range and occurs in most parts of the USA.

Cuscuta spp. belong to the Convolvulaceae family (same as the morning glory vine) but are yellow instead of green. Some species do have a small amount of chlorophyll but cannot support themselves and must live as parasites. Dodder is favored by high temperature and full sunlight.

Disease Cycle and Epidemiology

The dodder plant reproduces from seed that overwinters in the soil or as a contaminant in alfalfa seed lots. The rootless dodder seedling emerges from the soil as a slender yellow stem that rotates slowly counterclockwise. When a plant is contacted, the dodder twines about it and then develops haustoria that penetrate the vascular system, thereby obtaining food material from the host. The base of the dodder seedling then withers and dies, so that it loses contact with the soil and lives completely at the expense of the host. The dodder seedling dies within three weeks if a host is not contacted.

Dodder grows rapidly and produces many entwining stems that reach out to attack adjacent plants. During the growing season, clusters of minute white, pink, or yellowish flowers produce abundant seed. The seeds fall to the ground or are harvested with the alfalfa hay or seed (Fig. 68).

The dodder is spread with alfalfa seed, in hay, by farm workers and implements, in irrigation and surface

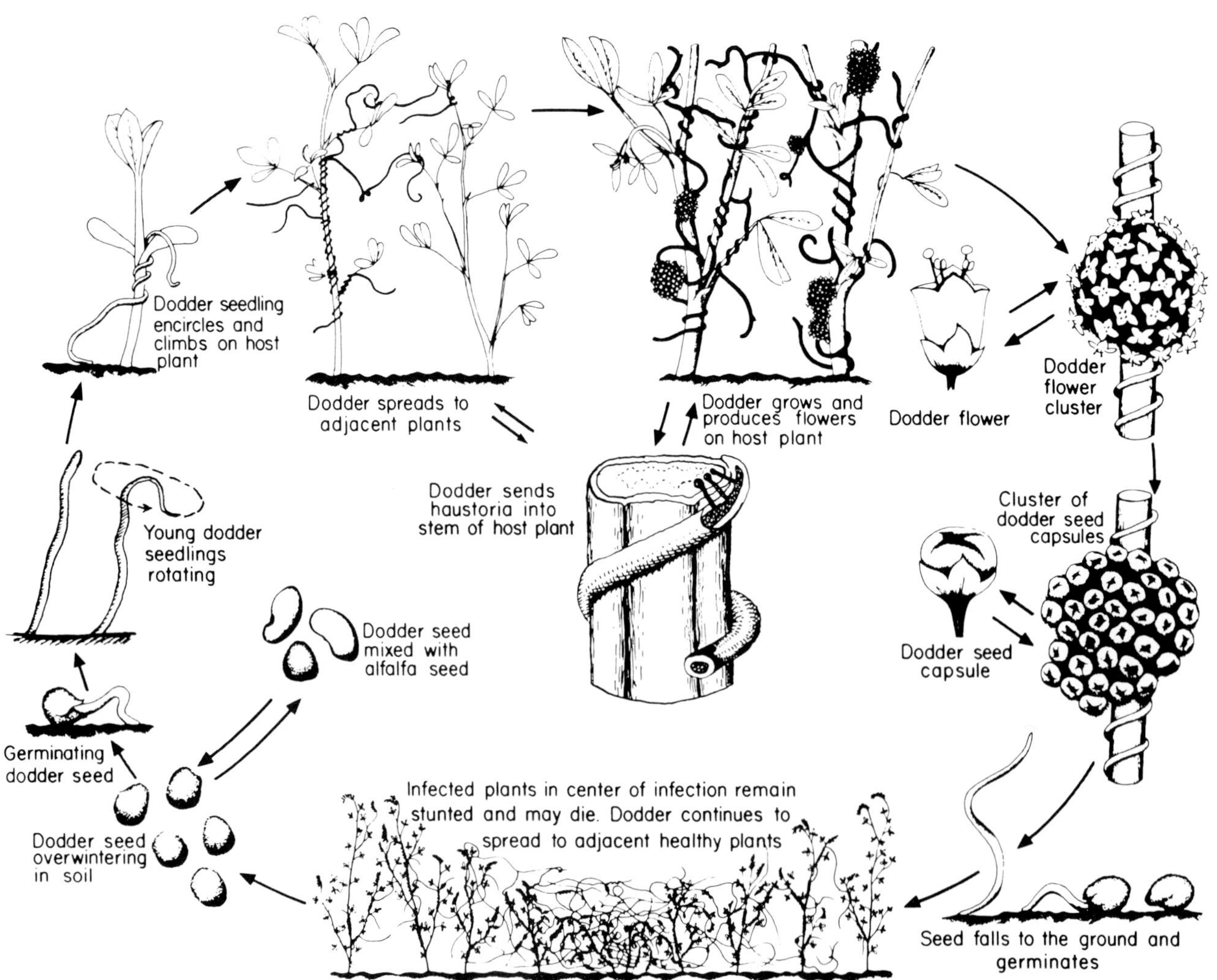

Figure 68. Life cycle of dodder (*Cuscuta* spp.). (Reprinted, by permission and with modification, from Plant Pathology by G. N. Agrios. ©1969, The Academic Press, New York)

drainage water, and in manure from animals fed contaminated hay. Weeds and other dicotyledonous crop plants are hosts and therefore can be sources of infestation.

Control

1) Prevention is the best control.

2) Use tagged, certified seed of known purity.

3) Small-seeded dodder can be removed from alfalfa seed by screening. A magnetic seed separator is used to separate large-seed dodder; fine iron powder clings to the rough-coated seed.

4) Use vigilance to discover and destroy any dodder in a field.

5) Chemical control in alfalfa seed fields depends on age of alfalfa and whether the crop is used for forage, amount of dodder in the field, feasibility of frequent cultivation, and character of field with respect to slope, soil, drainage, etc. A combination of treatments is usually required.

6) Control must be 100% effective.

Selected References

DAQSON, J. H., W. O. LEE, and F. G. TIMMONS. 1971. Controlling dodder in alfalfa. U.S. Dep. Agric. Farmers Bull. 2211. 16 p.

YUNCKER, T. G. 1932. The genus Cuscuta. Mem. Torrey Bot. Club. 18:113-331.

Insect Injury

Potato Leafhopper Burn

The potato leafhopper, *Empoasca fabae* (Harris), is an important pest on second and successive alfalfa crops in the eastern half of the USA. The potato leafhopper occurs in the Western hemisphere from Argentina to Canada.

The adult potato leafhopper is a light green, wedge-shaped, winged insect about 3 mm long and less than 1 mm wide at the head, tapering evenly to the wing tips.

Adults and nymphs feed on leaf veins and petioles, causing a characteristic V-shaped yellowing with occasional reddening or purpling beginning at the tip and progressing toward the base of the leaflet (Plate 47). Accompanying these symptoms are stunting of plants and reduced forage protein, carotene, and yield. Mechanical and enzymatic destruction of phloem tissues at the feeding site blocks the downward transport of photosynthate, resulting in an accumulation of anthocyanin. In the early stages of development, potato leafhopper-yellowing may appear much like symptoms of boron deficiency.

Control

Timely application of appropriate insecticides.

Selected Reference

JOHNSON, H. W. 1934. Nature of injury to forage legumes by the potato leafhopper Empoasca fabae (Harris). J. Agric. Res. 49:379-406.

(Prepared by J. L. Kugler)

Spotted Alfalfa Aphid Vein Clearing

The spotted alfalfa aphid, *Therioaphis maculata* (Buckton), was introduced into North America from the Near East in about 1953 and is now present in 38 states of the USA and in regions of Mexico and Australia. It is pale yellow, with six rows of black spots along its back, and is about 1–1.5 mm long. Both winged and wingless adult forms exist. Populations can increase rapidly by parthenogenesis. Adults and eggs may overwinter. Alternate hosts include several clovers. The spotted alfalfa aphid is smaller but more damging to alfalfa than either the pea aphid, *Acyrthosiphon pisum* (Harris), or the blue alfalfa aphid, *A. kondoi* (Shinji).

The aphid sucks juices from phloem of leaves and stems and often kills young alfalfa plants of susceptible cultivars. Susceptible mature plants are severely stunted and covered with a sticky honeydew secreted by the aphid. Various sooty mold fungi may colonize the honeydew.

Early plant symptoms include vein clearing (chlorosis) of newly emerging leaves, often within 48 hr

Figure 69. Vein clearing in expanding and young alfalfa leaves caused by salivary toxin from spotted alfalfa aphid *(Therioaphis maculata)* feeding. (Courtesy Department of Entomology, Kansas State University)

of continual aphid feeding. As feeding continues, veins may whiten before remaining areas of the leaf turn yellow. Eventually leaves die and fall from the plant. Stand life is reduced. The aphid produces a salivary toxin that is responsible for vein clearing, loss of leaves, and stunting and death of plants (Fig. 69).

Control

1) Plant resistant cultivars.

2) Use insecticides when the aphid population reaches an economically damaging level.

3) Biological control of the aphid by fungal and insect parasites is effective in many situations.

Selected References

DAVIS, D. W., NICHOLS, M. P., and ARMBRUST, E. J. 1974. The literature of arthropods associated with alfalfa. I. A bibliography of the spotted alfalfa aphid, Therioaphis maculata (Buckton) (Homoptera: Aphidae). Ill. Nat. Hist. Surv. Biol. Notes No. 87.

1969. The spotted aphid. How to control it. U.S. Dep. Agric. Leaflet No. 442. 8 p.

(Prepared by J. L. Kugler)

Stem Girdle

The three-cornered alfalfa hopper, *Stictocephala festina* (Say), is an insect that causes girdling of the lower stems of alfalfa and other plants in the irrigated desert valleys of the Southwest and in the southern states of the USA. Adults and nymphs suck plant juices by puncturing the stem, often in a regular line that completely girdles it (Fig. 70). This damage restricts movement of photosynthate from leaves to roots.

The leaves on affected stems become purple to yellow, and eventually the stem dies and breaks (Plate 48). The injury can be confused with some stem canker diseases caused by fungi.

There is no known control.

Figure 70. Girdled areas on alfalfa stems caused by the three-cornered alfalfa hopper *Stictocephala festina*. (Courtesy D. C. Erwin)

Selected Reference

WILDERMUTH, V. L. 1915. Three cornered alfalfa hopper. J. Agric. Res. 3:343-364.

Guide to the Identification of Biotic Diseases

Only a positive identification of the causal agent assures correct diagnosis of disease. Signs that indicate the causal organism are set in boldface type.

Foliar Diseases

Symptoms	Disease	Page	Figure	Plate
Only on frost-damaged spring growth. Rare in occurrence. Stem lesions begin at point of leaf attachment and extend down. Water-soaked lesions become amber-colored and shiny (bacterial exudate), then blacken with age. **Bacteria in ooze from water-soaked areas.**	Bacterial stem blight	15	...	...
Favored by low temperatures. Small irregular black to brown spots enlarge and coalesce on leaves. Infected leaves turn yellow, wither and fall. Stems blacken. **Pycnidia with conidia on overwintered stems.**	Spring black stem	17	...	17
Favored by cool moist conditions. Most severe on seedling stands. Yellow blotches on upper side of leaves; **white to purplish downy growth (conidiophores and conidia) on the lower surface**. Systemically infected shoots shortened and chlorotic; terminal leaves twisted.	Downy mildew	26	33	24
Favored by cool humid conditions. Leaves and stems become flaccid and collapse. **White fluffy mycelium grows over dead parts and produces hard black sclerotia on and in infected tissue.**	Sclerotinia crown and stem rot	33	40	30, 31
Most prevalent in late spring and early summer. Lesions elongate chlorotic to yellowish-orange blotches parallel to leaf veins. **Numerous dark pinpoint dots (pycnidia) develop on the blotched areas, especially on upper surface.**	Yellow leafblotch	20	...	18
Young regrowth "burned" (large bleached areas) during or shortly after cool rainy periods. On older leaves, 1–3 mm diameter brown eyspots with lighter centers and surrounded by chlorotic halos. **Ascocarps mostly on decayed leaves.**	Leptosphaerulina leaf spot	21	25	19
Lesions brown, 1–3 mm in diameter, finely serrated margins; rarely coalesce. **A raised disc (apothecium) to 1 mm in diameter develops in the center of mature lesions on the upper leaf surface.** Leaves turn yellow and drop.	Common leaf spot	19	23, 24	...

(continued on next page)

Foliar Diseases (continued)

Symptoms	Disease	Page	Figure	Plate
Favored by warm to hot humid conditions. Stem lesions sunken, mostly near base of plant, whitish to tan, oval to diamond shaped, brown borders; may girdle stems. **Black specks (acervuli) develop in lesions.**	Anthracnose	25	30, 31, 32	...
Bleached elongated lesions with brown bands that girdle stems at base of plant. **Coarse brown mycelium usually on lesion.**	Rhizoctonia stem blight	31	...	27
Favored by hot moist conditions. Leaf lesions roughly circular, reddish brown, to 6 mm diameter. **Conidia give lesions a silver appearance.** Brown stem lesions enlarge and coalesce.	Summer black stem	23	27	22
Favored by high temperatures. Leaf lesions irregular with yellowish to tan papery translucent centers and narrow water-soaked borders that are most pronounced on lower leaf surface. Stem lesions greasy, water-soaked, and become brown; usually coalesce. **Bacteria in ooze from water-soaked areas.**	Bacterial leaf spot	15	...	16
Favored by wet periods during summer and fall. Lesions oval, slightly sunken, dark brown with lighter centers and chlorotic halos. Older lesions may be concentrically ringed. **Dark conidia develop profusely in older lesions.**	Stemphylium leaf spot	21	26	20, 21
Bleached, roughly circular leaf spots to 1 cm diameter with brown borders and **dark specks (pycnidia).** May be associated with crown and root rot.	Stagonospora leafspot	24	28, 29	23
Most prevalent late in the growing season. **Circular reddish brown pustules protruding from leaves and stems. The "rusty spots" (spores) rub off easily.**	Rust	27	34	25

Crown and Root Diseases

Symptoms	Disease	Page	Figure	Plate
Seedlings either may not emerge from soil or may become constricted and rot off at the base of the stem. **Damping-off can be caused by Rhizoctonia solani (septate mycelium)** or by one or several *Pythium* spp. (**Nonseptate mycelium and/or oospores in affected tissue.**)	Damping-off	16	21	...
Dark bluish black necrotic area, sometimes V-shaped, extending into the crown from diseased stems (stubble).	Anthracnose crown rot	25	30, 31, 32	...
Foliage becomes yellow and flaccid. **Mycelium produced on crown is white and fluffy. After infected tissue is colonized, hard black sclerotia (with white interior) form.**	Sclerotinia crown and stem rot	33	40	30, 31
Black discoloration of lateral and taproots. **Black soft sclerotia form in decayed tissue.**	Mycoleptodiscus crown and root rot	39	49, 50, 51	...
Dark brown rotted tissue on the crown or on crown buds occurring under snow cover in patches in northern USA and in Canada. **White mycelium of the causal fungus (unidentified basidiomycete) can be seen on soil surface.** Fungus produces hydrogen cyanide.	Winter crown rot or snow mold	38	48	...

Crown and Root Diseases (continued)

Symptoms	Disease	Page	Figure	Plate
Plants are stunted with small, dark green leaves resembling drought stress. Xylem tissue in taproot is dark brown.	Dwarf	14	20	14, 15
Plants are stunted with small cupped leaves, yellow on margins; many crown shoots. Diseased plants scattered through stand. Xylem tissue in taproot is tan to brown; discoloration sometimes extends laterally to the inside surface of the bark. Usually does not appear before second or third crop year.	Bacterial wilt	13	19	12, 13
Plants wilt and the xylem tissue of the taproot becomes brown. Stem remains green for a while after all leaves on a shoot are killed.	Verticillium wilt	35	41, 42	33, 34
Plants usually with lower leaf yellowing, often a one-sided wilt. Longitudinal streaks in xylem tissue are brick red to dark brown. **Mycelium may be seen in xylem vessel elements.**	Fusarium wilt	36	43	35
Cortex and xylem tissue with orange to brick-red flecks. Very distinctive.	Stagonospora root and crown rot.	24	28, 29	23
Necrotic lesions on taproots are brown, sunken, and irregular in shape. Roots may be entirely rotted. **Rhizomorphs with cruciate hyphae on the surface are the most diagnostic sign**. Found only in southwestern USA and in northern and central Mexico.	Phymatotrichum root rot	37	44, 45, 46, 47	36, 47
Brown to yellow elliptical cankers with entire margins on the taproot in the summer at point of emergence of the lateral roots. Cankers become black and inactive during winter months. Disease is important only in irrigated high-temperature desert areas.	Rhizoctonia root canker	30	37	...
Occurs in saturated soils. Affected area on the taproot is brown to yellow, eventually turns black, with diffuse and irregular margins. Rotted tissue is usually firm. Taproot may rot off completely. **Oospores form in rotted tissue.** New roots may emerge above rotted areas after soil dries.	Phytophthora root rot	29	36	26
Plants in low spots in field die. Patches are circular to irregular and enlarge as disease progresses. **Violet mycelium eventually forms a violet felt on taproots**. Bark loosens and slips from the central cylinder.	Violet root rot	32	...	28, 29
Root rot characterized by brown sunken lesions occurs in Canada. **Brown to black pycnidia occur on surface of lesions or are embedded in the necrotic tissue**.	Plenodomus root rot	41	53	...
Light brown general root or crown rot extending either a short distance below the crown or rotting the entire root. **Sclerotialike stromata of the causal organism may develop in cracks in the affected cortex of the taproot**. Occurs in Canada and northern USA.	Cylindrocarpon root rot	40	52	38

(continued on next page)

Crown and Root Diseases (continued)

Symptoms	Disease	Page	Figure	Plate
Dark brown necrotic tissue in crowns and roots. Affected area may range from irregular rotted areas in cortex to complete disintegration of the root and crown.	Associated fungi include *Fusarium* spp., *Rhizoctonia solani,* and *Phoma* spp., combined with stress factors	42	54	39, 40

Glossary of Terms and Abbreviations

m—meter; 39.37 inches
cm—centimeter; 0.01 meter
mm—millimeter; 0.001 meter
μm—micrometer (micron); 0.000,001 meter (10^{-6}m)
nm—nanometer (millimicron); 0.000,000,001 meter (10^{-9} m)
in.—inch; 2.54 cm

acervulus (pl. acervuli)—an erumpent, saucer-shaped, cushionlike fruiting body of a fungus bearing condiophores, conidia, and sometimes setae
aeciospore—binucleate spore produced in an aecium of a rust
aecium—a cuplike sorus of the rusts
allantoid—somewhat curved with rounded ends; sausage-shaped
amphigynous—antheridium encircles the base or stalk of the oogonium
antheridium (pl. antheridia)—male sex organ (male gametangium) in fungi
apex—uppermost point
apothecium (pl. apothecia)—cuplike or saucerlike, ascus-bearing fruiting body (ascocarp) of Discomycetes, a group of fungi
appressorium (pl. appressoria)—swelling on a germ tube or hypha, especially for attachment to a host in early stage of infection
ascocarp—organ of a fungus that produces asci and ascospores
ascomycete—class of fungi in which the sexual spores (ascospores) are produced in a saclike structure, the ascus
ascospore—spore borne in an ascus
ascus (pl. asci)—saclike structure containing ascospores (typically eight) and borne in a fungus fruiting body
aseptate—without cross walls
asexual—vegetative, having no sexual organ or sexual spores; the imperfect stage of a fungus

bacilliform—shaped like short rods with rounded ends
bacteria—microscopic one-celled organisms that increase by binary fission (division of a mother cell into two like daughter cells)
basidiomycete—class of fungi characterized by basidia
basidiospore—sexual spore produced on a basidium
basidium (pl. basidia)—structure in basidiomycetes (includes smuts and rusts) on which basidiospores are borne

caespitose—in groups or tufts like grass
capsulate—enclosed in a capsule
capsule—gel-like material surrounding a bacterial cell
chlamydospore—thick-walled, asexual spore (terminal or intercalary) formed from hyphal cells or by transformation of conidial cells
chlorosis (adj. chlorotic)—disease symptom in higher plants; leaves or other green parts become yellow or blanched
clamp connection—hyphal outgrowth that, at cell division, makes a connection between the resulting two cells by fusion with the lower
clavate—club-shaped
coalesce—grow together into one body or spot
coenocytic—nonseptate; nuclei not separated by cross walls
concentric—one circle within another with a common center
conidiophore—simple or branched hypha on which conidia are produced
conidium (pl. conidia) **conidiospore**—any asexual spore borne at the tip or side of a specialized hypha, called a conidiophore
cruciate—cross-shaped (cruciform)
cuticle—outermost noncellular layer of plant tissue or nematodes

dentate—toothed
dichotomous—branching, frequently successively, into two approximately equal arms
distal—far from the point of attachment or origin

echinulate—spores having small spines or pointed processes projecting from cell walls
ellipsoid—surface, all plane sections of which are ellipses
elliptic—shaped like an ellipse
equatorial—relating to the equator or an equator
exudate—liquid discharge from diseased tissue or roots

falcate—sickle shaped
filamentous, filiform—threadlike
flaccid—limp, flabby
flagellum (pl. flagella)—whiplike appendage of a motile spore or bacterium
floccose—cottony
Fungi imperfecti (Deuteromycetes)—group of fungi for which the sexual stage is not known or the asexual stage of ascomycetes
fungus (pl. fungi)—organism having no chlorophyll, reproduction by sexual or asexual spores and not by fission, and, generally, a mycelium with well-marked nuclei
fusiform—spindlelike, narrowing toward the ends

gall—swelling or outgrowth produced by a plant as the result of attack by a fungus or other organism
geniculate—bent like a knee
globoid, **-ose**, **-ular**, **-ulose**—almost spherical
Gram-negative—bacteria staining red after treatment with Gram's stain
Gram-positive—bacteria staining violet after treatment with Gram's stain
gravid—containing eggs, capable of depositing eggs

haustorium (pl. haustoria)—special hyphal branch or outgrowth of a parasitic plant, especially one in a host cell for absorption of food
hermaphrodite (adj. hermaphroditic)—having both male and female reproductive organs
homothallic—condition in which a haploid thallus is self-fertile
hyaline—colorless, transparent
hydathode—epidermal structure functioning in the exudation of water
hymenium—spore-bearing layer of a fungus fruiting body
hypertrophy—abnormal increase in the number of cells, resulting in formation of galls or tumors
hypha (pl. hyphae)—thread of the fungus mycelium

imperfect (stage)—asexual spores (such as conidia) or no spores are produced
imperfect fungi—see Fungi imperfecti
infection—process of beginning or inducing disease
inoculum—substance, generally a pathogen, used to produce disease or one that produces disease in nature

inoperculate—having no operculum (cover or lid)
inter- (prefix)—between, among; -calary cell, a cell between two others; -cellular, between host cells
intra- (prefix)—within, inside; -cellular, in a cell

latent—present but not visible
lesion—well marked but limited disease area; a wound
lobate—lobed
lobulate—having small lobes
locule—cavity, especially one in a stroma

macroconidium (pl. macroconidia)—the larger, generally more diagnostic conidium of a fungus (e.g. *Fusarium*) that also has microconidia; a long or large conidium
matrix—substratum in or on which a fungus lives
meristem—formative plant tissue (growing point)
microconidium (pl. microconidia)—the smaller conidium of a fungus that also has macroconidia; a small conidium often acting as a spermatium
muriform—having cross and longitudinal septa
Mycelia sterilia—Fungi imperfecti that have no spores, except in some cases, chlamydospores
mycelium (pl. mycelia)—mass of hyphae constituting the body (thallus) of a fungus

necrosis (adj. necrotic)—death of plant cells, usually resulting in tissue darkening; commonly a symptom of fungus, nematode, virus, or bacterial action
nematode—small wormlike animal, parasitic in plants or animals, or free living in soil or water
nucleus—portion of cell enclosed in a membrane in which phenomena and heredity are controlled

obligate parasite—parasite that can develop only in living tissue
oogonium (pl. oogonia)—female sex organ of oomycetes, a large group of fungi
oosphere—female gamete in oomycetes
oospore—thick-walled, resting spore that develops from an oosphere produced by fertilization or parthenogenesis
ostiole—pore in a perithecium or pycnidium
ovate, **ovoid**—egg-shaped

papilla (pl. papillae)—small rounded or nipplelike projection; (in *Phytophthora*) a plug in the tip of the sporangium
papillate—having a papilla
paragynous—antheridium contacts oogonium laterally above its base
paraphysis (pl. paraphyses)—sterile, upward growing hyphal element in a hymenium, especially in ascomycetes; usually clavate or filiform
parasite—organism living in or on another organism in parasitism
parenchyma—variable soft tissue composed of approximately isodiametric living cells with thin cell walls and large vacuoles; often with plastids
parthenogenesis (adj. parthenogenetic)—reproduction by the development of an unfertilized egg
pathogen—organism or agent capable of causing disease
perfect (stage)—sexual spores (oospores, zygospores, ascospores, or basidiospores) are formed, generally after nuclear fusion
pH—measure of acidity. pH 7 is neutral; below pH 7, acidic; above pH 7, alkaline
phloem—complex vascular tissue including sieve tubes (through which food moves), companion cells, fibers, and parenchyma
pionnotes—spore mass having a greaselike appearance (of *Fusarium*)
plano-convex—convex but somewhat flat
pleomorphic—occurrence of more than one distinct form in the life cycle of an organism
plerome (procambium strands)—inner tissue in root meristem
procaryotic—lacking a nuclear membrane, mitotic apparatus and mitochondria; antonym: eucaryotic
protrusile—capable of being protruded
pustule—blisterlike, frequently erumpent spot or spore mass
pycnidiospore—conidium in or from a pycnidium
pycnidium (pl. pycnidia)—asexual, flask-shaped or globose fungus fruiting body containing conidia
pycniospore—spore formed in a pycnium of a rust fungus; also called a spermatium
pycnium—haploid fruiting body (in rusts); like a pycnidium

rickettsia—pleomorphic, rod-shaped, nonfilterable microorganism that causes disease (especially animal)

saccate—like a sac or bag
saprophyte—organism that uses dead organic matter as food
sclerotium (pl. sclerotia)—hard, frequently rounded and usually darkly pigmented resting body of a fungus composed of a mass of special hyphal cells. The structure may remain dormant for long periods; when favorable conditions return,sclerotia germinate to produce a stroma, fruiting body, mycelium, or conidiophores
septate—having septa
septum (pl. septa)—cross wall in a hypha
serrate—edges with teeth, like a saw
seta (pl. setae)—stiff hair, usually dark and thick-walled
sporangiophore—hypha bearing sporangia
sporangium (pl. sporangia)—saclike fungal structure in which the entire contents are converted into an indefinite number of asexual spores
spore—single-celled or many-celled reproductive body in the fungi and lower plants that can develop into a new plant
sporodochium (pl. sporodochia)—cushion-shaped stroma covered with a mass of conidiophores
sporulate—to produce spores
stele—central cylinder of vascular tissue (especially in roots)
sterigma (pl. sterigmata)—process from a cell supporting a spore
stroma (pl. stromata)—compact mass of specialized hyphae, sometimes sclerotiumlike in form, in or on which spores are produced
submicroscopic—too small to be seen in an ordinary light microscope

teliospore—thick-walled resting spore produced by some fungi, notably rusts and smuts, from which the basidium is produced
thallus—vegetative body of a fungus
tolerant—refers to plants capable of sustaining disease without serious injury or crop loss
toxin—poison produced by an organism
tracheid—elongated conducting cell of the xylem, with tapering or oblique end walls and pitted walls
truncate—ending abruptly as though the end had been cut

urediospore—binucleate, repeating spore of rust fungi
uredium (pl. uredia)—sorus bearing urediospores

vector—agent, usually an insect, that carries and transmits disease-causing viruses, bacteria, fungi, or mycoplasma
verticillate—whorled
vessel—water-conducting cell usually with open end walls in xylem tissue
viruliferous—virus carrying, especially virus-laden insects
virus—submicroscopic, filterable agents that cause disease and multiply only in living cells

xylem—water-conducting tissue in plants

zoospore—motile, asexually produced fungus spore having flagella, a swarm spore

Index